AQA
GCSE Science
CORE FOUNDATION

Editor: Graham Hill

Nigel Heslop, Graham Hill,
Toby Houghton, Steve Witney
and Christine Woodward

Hodder Murray

A MEMBER OF THE HODDER HEADLINE GROUP

The Publishers would like to thank the following for permission to reproduce copyright material:
p.9 Science Photo Library/Saturn Stills; **p.10** Corbis/Sean Aidan/Eye Ubiquitous; **p.11** *t* Still Pictures/Michael J. Balick, *b* Corbis/Eric and David Hosking; **p.12** *t* Science Photo Library/AJ Photo, *b* Rex Features/Mike Webster; **p.14** Bob Battersby; **p.18** Anthony Blake Photo Library/Maximilian Stock Ltd.; **p.22** *l* Anthony Blake Photo Library/Joy Skipper, *c* Anthony Blake Photo Library/Maximilian Stock Ltd, *r* Bob Battersby; **p.23** Science Photo Library/John Radcliffe Hospital; **p.24** *l* Rex Features/Henryk T. Kaiser, *c* Rex Features/John Powell, *r* Rex Features/Image Source; **p.25** *t* Corbis/Norbert Schaefer, *c* Rex Features/Image Source, *r* Alamy/Alaska Stock LLC; **p.30** Rex Features/TS/Keystone USA; **p.32** *t* Corbis/Visuals Unlimited, *b* Corbis/Mediscan; **p.33** Science Photo Library; **p.35** Rex Features/Garo/Phanie; **p.38** Science Photo Library/Nancy Sefton; **p.39** *t* Rex Features/John W. Warden, *c* Photolibrary/OSF/David Cayless, *b* Photolibrary/OSF/Stephen Shepherd; **p.40** *tl* Science Photo Library/John Beatty, *tr* Bruce Coleman Ltd, *bl* Ardea/Bob Gibbons; **p.41** *l* Rex Features/Reso, *r* Alamy/John Sylvester; **p.43** *t* Alamy/Brian Elliott, *b* Science Photo Library/Dr Jeremy Burgess; **p.44** Photodisc; **p.45** *t* Still Pictures/Mark Edwards, *l* Photolibrary/OSF/Harold Taylor, *cl* Photolibrary/OSF/Paulo De Oliveira, *cr* Still Pictures/Luc Vausort, *r* Science Photo Library/David Aubrey; **p.46** Alamy/Worldwide Picture Library; **p.50** *t* Getty Images/David Woodfall, *bl* Christine Woodward, *br* Christine Woodward; **p.52** Getty Images/Art Wolfe; **p.54** Christine Woodward; **p.56** *t* Christine Woodward, *b* Alamy/Jim West; **p.60** *tl* Science Photo Library/CAMR/A.B.Dowsett, *tr* Corbis/Stuart Westmorland, *bl* Rex Features/Richard Austin, *br* Rex Features/Rex Interstock; **p.61** Science Photo Library/James King Holmes; **p.64** *t* Science Photo Library/Kenneth W. Fink, *b* Science Photo Library/SCIMAT; **p.65** Corbis/Ian Harwood/Ecoscene; **p.67** Science Photo Library/Michael Donne, *r* Still Pictures/Philippe Hays; **p.69** *tl* Still Pictures/Martin Harvey, *tr* Science Photo Library/Gregory Dimijian, *c* Photolibrary/OSF/Sue Scott, *br* Mary Evans Picture Library; **p.70** Corbis/Leonard de Selva; **p.71** Hulton Archive/Getty Images; **p.72** Corbis/Reuters; **p.73** Science Photo Library/George Bernard; **p.76** Stephanie Maze/CORBIS; **p.77** *t* Rex Features/Nils Jorgensen, *b* Corbis/Chris Bland/Eye Ubiquitous; **p.80** Geoscience Features Picture Library; **p.85** Rex Features/Dave Penman; **p.87** *t* Corbis/Tim McGuire, *b* Rex Features/Sunset; **p.88** *tl* Russ Merne/Alamy, *cl* Danita Delimont/Alamy, *cr* Greenshoots Communications/Alamy; **p.90** *cl* Philadelphia Museum of Art/CORBIS, *c* Photolibrary.com, *cr* Rob Melnychuk/Taxi/Getty Images; **p.92** Geoscience Features Picture Library; **p.93** Steve Atkins/Alamy; **p.96** Peter Bowater/Alamy; **p.97** *t* Pascal Goetgheluck/Science Photo Library, *b* Scott Camazine/Alamy; **p.103** Getty Images/Hulton Archive; **p.105** Nigel Heslop; **p.106** Nigel Heslop; **p.108** Nigel Heslop; **p.112** Still Pictures/Ray Pfortner; **p.114** *t* Corbis/Peter Turnley, *bl* Nigel Heslop, *br* Nigel Heslop; **p.115** Bob Battersby; **p.116** Nigel Heslop; **p.117** *t* Geoscience Features, *b* Nigel Heslop; **p.119** Bob Battersby; **p.120** *tl* Photodisc, *cl* Nigel Heslop, *c* Nigel Heslop, *cr* Nigel Heslop, *bl* Nigel Heslop; **p.124** *t* Getty Images/UHB Trust, *b* Still Pictures/Wolfgang Maria Weber; **p.125** Science Photo Library/Jerry Mason; **p.128** Ingram; **p.129** Ingram; **p.130** *tc* Photodisc, *tr* Rex Features/Burger/Phanie, *cl* Bob Battersby; **p.131** *t* Anthony Blake Photo Library/Tim Hill, *b* Bob Battersby; **p.132** Alamy/Marie-Louise Avery; **p.133** *t* Andrew Lambert, *c* Science Photo Library/Maximilian Stock Ltd., *b* Rex Features/The Travel Library; **p.136** Alamy/AGStockUSA, Inc; **p.140** Science Photo Library/Planetary Visions Ltd.; **p.141** *t* Corbis/Jim Craigmyle, *c* Getty Images/Wayne Levin; **p.148** Rex Features/RYB; **p.150** GeoScience Features; **p.151** *c* Getty Images/AFP, *b* Rex Features/Sipa Press; **p.154** Rex Features/Sipa Press; **p.155** Rex Features/The Travel Library; **p.159** *cl* Ecoscene/Bruce Harber, *cr* Courtesy Sony, *bl* Rex Features/Sipa Press, *br* Still Pictures/Hartmut Schwarzbach; **p.161** *tl* Science Photo Library/Gusto, *bl* Rex Features/Shout; **p.165** Rex Features/David Cole; **p.168** Corbis/Walter Hodges; **p.172** *c* Corbis/Dimitri Iundt, *b* Rex Features/Phil Ball; **p.174** Lorna Ainger; **p.178** *tl* Corbis/Paul A. Souders, *tr* Still Pictures/David Woodfall/WWI, *cl* Science Photo Library/Tony Craddock, *cr* Science Photo Library/Martin Bond, *bl* Rex Features/Reso, *br* Still Pictures/Jim Wark; **p.181** *cl* Bob Battersby, *cr* Andrew Lambert, *bl* Getty Images/Luc Hautecoeur, *br* Getty Images/Barry Rosenthal; **p.184** Rex Features/BYB; **p.189** Science Photo Library/Martin Bond; **p.190** *t* Corbis/Richard Cummins, *b* Science Photo Library; **p.192** *t* Courtesy LVM/Aerogen Wind Generators, *b* Courtesy NPower; **p.193** Still Pictures/Jorgen Schytte; **p.194** *t* Rex Features/Reso, *c* courtesy of Better Energy Systems Ltd./Solio™, *b* Bob Battersby; **p.195** Rex Features/The Travel Library; **p.198** *l* Science Photo Library/Martin Dohrn, *tc* Still Pictures/Jorgen Schytte, *bc* Rex Features/Tess Peni, *tr* Science Photo Library/Zephyr, *cr* Rex Features/Roy Garner; **p.199** *t* Empics/Steve Mitchell, *b* Corbis/Craig Tuttle; **p.200** Rex Features/Sipa Press; **p.204** *t* Getty Images/Katsumi Suzuki, *c* Alamy/Gary Pearl, *b* Rex Features/Eye Ubiquitous; **p.205** Science Photo Library/Klaus Guldbrandsen; **p.206** *t* Science Photo Library, *b* Science Photo Library/Jim Varney; **p.207** Science Photo Library/Larry Mulvehill; **p.208** Andrew Ward; **p.211** *t* Science Photo Library, *b* Science Photo Library; **p.212** Rex Features/Image Source; **p.214** Corbis/Underwood & Underwood; **p.217** Science Photo Library/Pascal Goetgheluck **p.218** Science Photo Library/Will McIntyre; **p.220** Science Photo Library/ISM; **p.223** NASA; **p.227** Science Photo Library/Simon Fraser; **p.229** Corbis/Joseph Sohm/Visions of America; **p.230** Science Photo Library/Tony & Daphne Hallas; **p.231** Science Photo Library/Robert Gendler.

b = bottom, *c* = centre, *l* = left, *r* = right, *t* = top

Acknowledgements
Waste Online for permission to use the pie chart data on page 122.

Every effort has been made to trace all copyright holders, but if any have been inadvertently overlooked the Publishers will be pleased to make the necessary arrangements at the first opportunity.

Although every effort has been made to ensure that website addresses are correct at time of going to press, Hodder Murray cannot be held responsible for the content of any website mentioned in this book. It is sometimes possible to find a relocated web page by typing in the address of the home page for a website in the URL window of your browser.

Risk Assessment
As a service to users, a risk assessment for this text has been carried out by CLEAPSS and is available on request to the publishers. However, the publishers accept no legal responsibility on any issue arising from this risk assessment: whilst every effort has been made to check the instructions of practical work in this book, it is still the duty and legal obligation of schools to carry out their own risk assessments.

Hodder Headline's policy is to use papers that are natural, renewable and recyclable products and made from wood grown in sustainable forests. The logging and manufacturing processes are expected to conform to the environmental regulations of the country of origin.

Orders: please contact Bookpoint Ltd, 130 Milton Park, Abingdon, Oxon OX14 4SB. Telephone: (44) 01235 827720. Fax: (44) 01235 400454. Lines are open 9.00–5.00pm, Monday to Saturday, with a 24-hour message answering service. Visit our website at www.hoddereducation.co.uk.

© Nigel Heslop, Graham Hill, Toby Houghton, Steve Witney, Christine Woodward 2006
First published in 2006 by
Hodder Murray, an imprint of Hodder Education,
a member of the Hodder Headline Group
338 Euston Road
London NW1 3BH

Impression number   10 9 8 7 6 5 4 3 2
Year   2011 2010 2009 2008 2007 2006

Cover photos Science Photo Library: dragonfly, Andy Harmer; house, Ted Kinsman; limestone, Alfred Pasieka.
Typeset in Times 11.5pt by Fakenham Photosetting Limited, Fakenham, Norfolk
Printed in Italy

A catalogue record for this title is available from the British Library

ISBN-10: 0 340 90708 8
ISBN-13: 978 0 340 90708 5

Contents

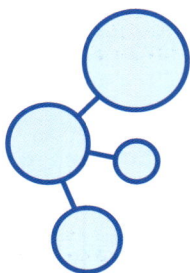

Introduction

AQA GCSE Science Core Foundation Student's Book

Welcome to the AQA GCSE Science Core Student's Book for the Foundation Tier. This book has been written to support you through your first Science GCSE with AQA. The book covers all the Biology, Chemistry and Physics material as well as the key 'How science works' elements of both the new specifications A and B.

Each chapter begins with a list of **learning objectives**. Don't forget to refer back to these when checking whether you have understood the material covered in a particular chapter. **Questions** appear throughout each chapter, which will help to test your knowledge and understanding of the subject, as you go along. They will also help you to develop key skills and understand how science works.

Activities are found throughout the book. These will take you longer to complete than the questions but will show you many of the real-life applications and implications of science. At the end of each chapter a **summary** provides the important points and key words. You will find the summaries useful in reviewing the work you have completed and in revising for your examinations. Don't forget to use the **index** to help you find the topic you are working on.

You will find **exam questions** at the end of each chapter, to help you prepare for your exams. These include similar questions to those in the module tests for Specification A and others like the structured questions for Specification B.

If you are studying Specification A with AQA you will take six tests. You will need to revise:

- Biology 1a Human Biology – chapters 1 and 2
- Biology 1b Evolution and Environment – chapters 3 and 4
- Chemistry 1a Products from Rocks – chapters 5 and 6 (sections 1 to 4)
- Chemistry 1b Oils, Earth and Atmosphere – chapters 6 (sections 5 to 9), 7 and 8
- Physics 1a Energy and Electricity – chapters 9 and 10
- Physics 1b Radiation and the Universe – chapters 11 and 12

If you are studying Specification B with AQA you will take three written papers, one for Biology (chapters 1 to 4), one for Chemistry (chapters 5 to 8) and one for Physics (chapters 9 to 12).

This book is written for the **foundation-tier** examination. It will push you so that you get the best possible result at GCSE. If you are planning to take the **higher-tier** examination, you may wish to use the Higher Student's Book instead.

Good luck with your studies!

Nigel Heslop, Graham Hill, Toby Houghton, Steve Witney and Christine Woodward

Chapter 1
How do our bodies respond to change?

At the end of this chapter you should:

✓ know how the different parts of your nervous system work;

✓ be able to explain the difference between a voluntary and a reflex response;

✓ understand that many processes in the body are controlled by chemicals called hormones;

✓ be able to explain how hormones control the amount of water and sugar in the blood;

✓ be able to describe how hormones control the menstrual cycle;

✓ know how hormones can be used to control a woman's fertility;

✓ be able to explain how medical drugs are developed and tested;

✓ be able to describe how drugs can affect our bodies;

✓ be able to consider and make decisions about the risks of taking cannabis.

Figure 1.1 In each of these pictures the body responds to what is happening.

1.1 How do different parts of our nervous system respond to changes?

Your body is always responding to changes. If someone throws you a ball you respond by moving your hands to catch it. When a teacher asks you a question, you may feel nervous. If you haven't drunk enough water on a hot day, your body will respond by trying to save the water you already have inside you.

In this section, you will learn how the **nervous system** controls your body's responses.

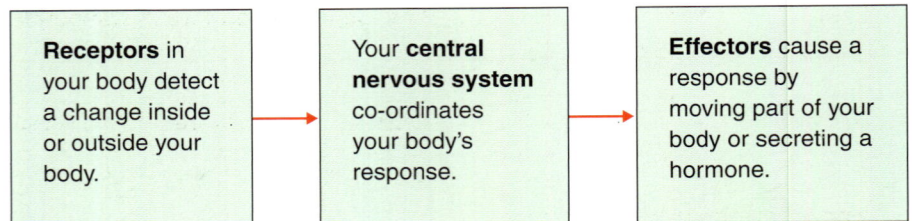

Look at the pictures in Figure 1.1, then copy and complete these sentences.

❶ If you touch a hot pan you respond by _____.

❷ If you wear a t-shirt in the cold your body responds by _____.

Receptors are cells in the body that can detect changes.

The **central nervous system (CNS)** is the spinal cord and the brain.

Effectors are organs in the body that cause a response. They are muscles or glands.

Sense organs are organs that contain receptor cells.

Stimuli (singular: **stimulus**) are changes that receptor cells detect.

Receptors in your body detect a change inside or outside your body.	Your **central nervous system** co-ordinates your body's response.	**Effectors** cause a response by moving part of your body or secreting a hormone.

Figure 1.2 A flow diagram showing how your nervous system enables your body to respond to changes

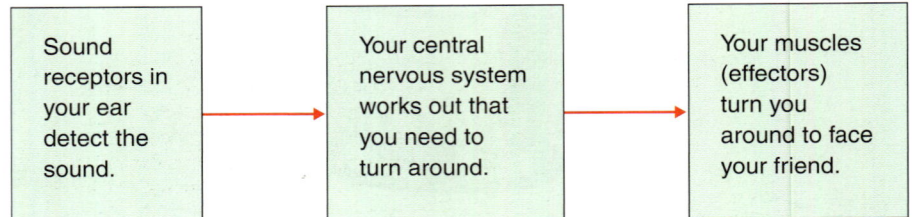

Sound receptors in your ear detect the sound.	Your central nervous system works out that you need to turn around.	Your muscles (effectors) turn you around to face your friend.

Figure 1.3 A flow diagram showing how you respond when a friend calls out your name

❸ Copy and complete the paragraph below, using the words in this box to fill in the blanks.

> effectors   eyes   nervous   receptors   response

When someone shines a torch at you, light _____ in your _____ detect the light. They send a message to your central _____ system. This sends a message to muscles, which are _____. Your muscles then cause a _____, which is to cover your eyes with your hand.

❹ a) Write down two changes that your body can respond to, other than light.

    b) Describe how the body responds to one of these changes.

Receptors are found in your **sense organs**. Eyes are the sense organs that contain light receptors. The changes that receptors detect are called **stimuli**. For example, light is the stimulus detected by the light receptors in your eyes. A receptor converts the stimulus into a tiny electrical impulse.

There are eight different types of receptors in your body. They are shown in Table 1.1.

Stimulus	Receptor	Sense organ
Light	Light receptors	Eye
Sound	Sound receptors	Ear
Movement	Position receptors Touch receptors Pressure receptors	Ear Skin Skin
Chemicals	Chemical receptors	Tongue Nose Blood vessels
Change in temperature	Temperature receptors	Skin Blood vessels Brain
Pressure / temperature	Pain receptors	Skin

Table 1.1 The types of receptors found in the body and the stimuli they detect

❺ Look at Figure 1.5.

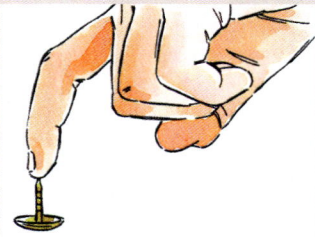

Figure 1.5 This person is about to respond to stimuli she can detect.

Use the words in the following box to help you fill in the gaps in Figure 1.6, which shows how the nervous system responds to touching a sharp drawing pin.

central	contract
electrical	impulse
muscles	nervous
receptors	response
skin	system

You should now be able to explain how the nervous system responds to a range of stimuli. A good way to do this is to draw a flow diagram. Use a box to describe each stage. Figure 1.4 shows how the nervous system responds when you touch a hot object, like a pan on a stove.

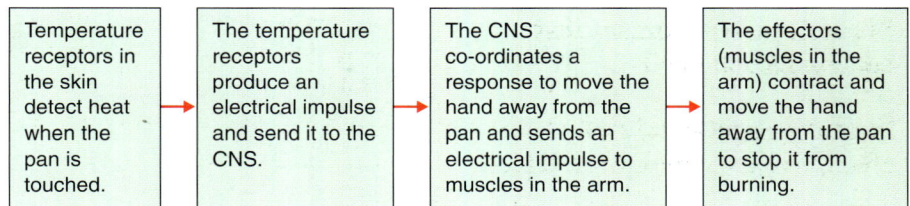

Temperature receptors in the skin detect heat when the pan is touched.	The temperature receptors produce an electrical impulse and send it to the CNS.	The CNS co-ordinates a response to move the hand away from the pan and sends an electrical impulse to muscles in the arm.	The effectors (muscles in the arm) contract and move the hand away from the pan to stop it from burning.

Figure 1.4 A flow diagram showing how the nervous system responds so that a person will not burn themselves seriously when they touch a hot object

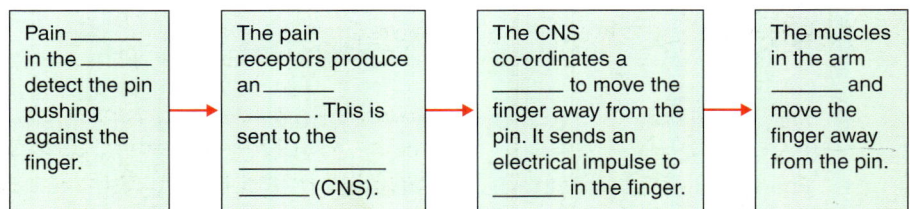

Pain _____ in the _____ detect the pin pushing against the finger.	The pain receptors produce an _____ _____ . This is sent to the _____ _____ _____ (CNS).	The CNS co-ordinates a _____ to move the finger away from the pin. It sends an electrical impulse to _____ in the finger.	The muscles in the arm _____ and move the finger away from the pin.

Figure 1.6 How the nervous system responds to a person touching a sharp drawing pin

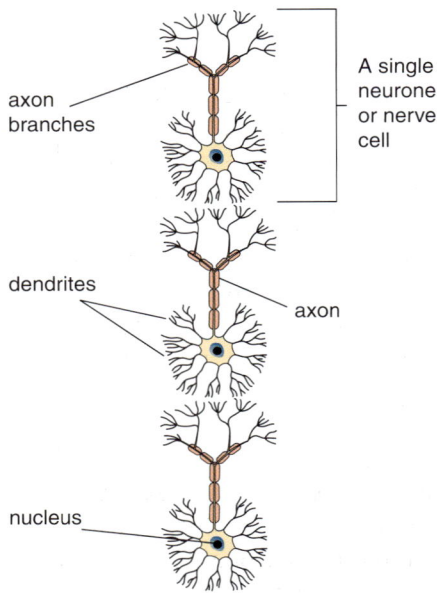

axon
branches

A single
neurone
or nerve
cell

dendrites

axon

nucleus

Figure 1.7 Three neurones forming one strand in a nerve

How do electrical impulses get around your body?

If you call a friend who lives two miles away, the phone lines connect you with your friend and carry your message to their phone. In a similar way, **nerves** in your body act like the wires in the phone lines. They conduct electrical impulses from one area to another, connecting the central nervous system to the receptors and the effectors.

Nerves are made up of individual nerve cells called **neurones**. Neurones are laid end to end to make long strands (Figure 1.7). These long strands are bundled together to make nerves (Figure 1.8). An electrical impulse starts at one end of a neurone and passes along it to the other end.

Between the end of one neurone and the next neurone there is a gap called a **synapse**. The electrical impulse cannot travel across this gap, but it causes a chemical to form (Figure 1.9). This **chemical transmitter** triggers an electrical impulse in the next neurone. Once the new impulse has been sent, the chemical is broken down by enzymes.

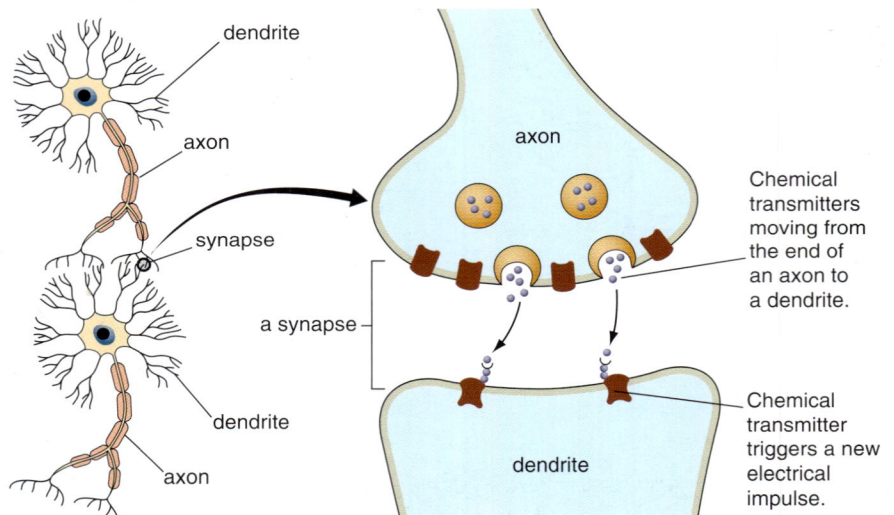

a nerve is made up of
a bundle of neurones

myelin sheath

individual
neurones

Figure 1.8 A bundle of neurones forming a nerve

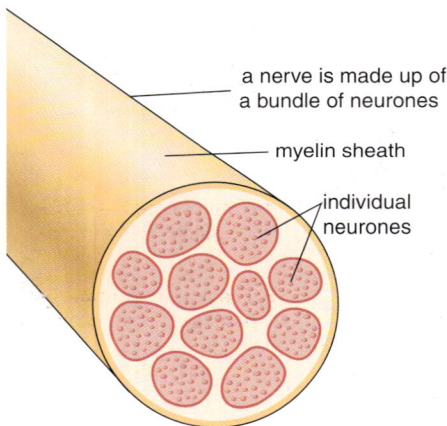

dendrite

axon

synapse

a synapse

dendrite

axon

axon

Chemical
transmitters
moving from
the end of
an axon to
a dendrite.

Chemical
transmitter
triggers a new
electrical
impulse.

dendrite

Figure 1.9 A chemical is released from one neurone. This travels across a synapse and triggers a new electrical impulse in the next neurone.

6 The seven sentences below describe how a signal travels from a temperature receptor in your finger to your brain. Write out the sentences in the correct order.
- A new electrical impulse is sent along the next neurone.
- The signal travels like this along your arm to the spinal cord.
- The temperature receptor sends an electrical impulse along a neurone.
- The impulse reaches the end of the neurone.
- A temperature receptor in your finger detects heat.
- The signal then travels up your spinal cord to your brain.
- A chemical passes across the gap (synapse) at the end of this neurone to the next neurone.

Neurones are individual nerve cells that carry electrical impulses.

Nerves are bundles of neurones connecting receptors and effectors to the central nervous system.

A **synapse** is the gap between the end of one neurone and the start of the next.

Chemical transmitters pass across a synapse and cause an electrical impulse in the next neurone.

Voluntary actions are responses that are co-ordinated by the brain.

Reflex actions are rapid, automatic responses, often to harmful situations.

Sensory neurones carry impulses from a receptor to the spinal cord.

Relay neurones carry impulses from sensory neurones to motor neurones.

Motor neurones carry impulses to effectors.

Figure 1.10 The reflex action allows a person to respond very quickly when they touch a sharp object.

How fast can the nervous system respond to stimuli?

Electrical impulses can travel very fast. They can move about 100 metres in one second. This makes your responses fast. But sometimes your brain takes time to work out how to respond to a signal. Responses that your brain co-ordinates in this way are called **voluntary actions**.

Voluntary actions are fine if you are responding to a question from your friend. But you need to respond more quickly to a harmful situation. For example, if you touch a sharp object, you need to pull your hand away without thinking about it. This type of sudden, automatic response is called a **reflex action**.

Reflex actions do not rely on the brain to work out how to respond. The spinal cord processes the response. A **sensory neurone** carries the impulse to the spinal cord. A **relay neurone** in the spinal cord passes the impulse straight to a **motor neurone**. The motor neurone then carries the impulse to the effector. The effector then responds.

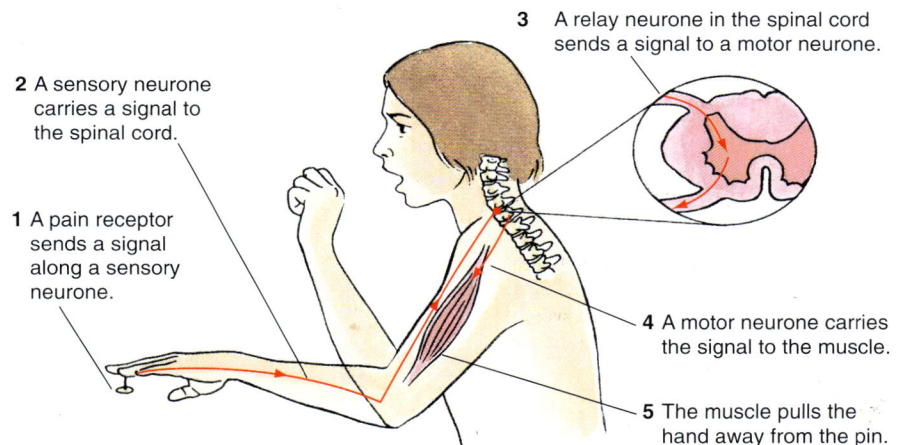

3 A relay neurone in the spinal cord sends a signal to a motor neurone.

2 A sensory neurone carries a signal to the spinal cord.

1 A pain receptor sends a signal along a sensory neurone.

4 A motor neurone carries the signal to the muscle.

5 The muscle pulls the hand away from the pin.

❼ Which of the responses listed below are voluntary actions and which are reflex actions?
- Blinking when someone kicks dust at you.
- Sneezing when pepper goes up your nose.
- Picking up a pen.
- Pulling your hand away when you touch a hot iron.
- Turning around when someone calls your name.
- Taking a CD off a shelf.

❽ Choose one of the reflex actions from question 7. Complete the sentences below to explain how the body responds.

_____ receptors detect _____.
↓
The receptors send an electrical impulse along a _____ neurone.
↓
The impulse then goes along a _____ neurone in the _____ cord.
↓
The impulse then goes along a _____ neurone.
↓
Finally the impulse reaches an _____.
↓
This causes _____.

Activity – Multiple sclerosis

Multiple sclerosis (MS) is a disease that damages the central nervous system. It can cause blindness, slurred speech, poor co-ordination, paralysis and memory loss. Two people with MS may suffer in very different ways. At present, there is no cure for multiple sclerosis. But scientists are developing treatments for the disease.

Nerves are coated with a chemical called myelin. This insulates the nerve like the plastic coating on electrical wire. A person with MS has damaged patches of myelin. This leaves the nerve exposed.

❶ What problems can people with MS have?
❷ How does damage to myelin cause these problems?
❸ MS damages only nerves in the brain and spinal cord. But a sufferer may have poor co-ordination in his/her legs. Why?
(Remember, nerves in the legs are not actually damaged.)
❹ Do you think that MS can be passed from one person to another? Explain your answer.
❺ When people realise they have MS they need to know more about the disease.
Write down six questions that these people might ask.

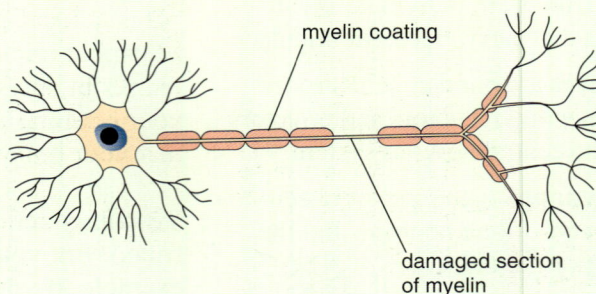

Figure 1.11 A nerve with a damaged myelin coating, like those in a person with MS

The following websites will help you with this question:
www.mstrust.org.uk
www.mult-sclerosis.org
❻ Scientists are still finding out new things. By developing our scientific knowledge, scientists have helped us to answer many questions. But science cannot answer all questions.
Look carefully at the six questions you have listed for question 5.
  a) Which of these questions do you think can be answered using our present scientific knowledge?
  b) Which of these questions are outside the boundaries of science? They may be questions where beliefs, opinions or personal views are important.

1.2

How do hormones control the conditions inside your body?

Hormones are chemicals. They travel in the blood to their target organ.

Glands are organs that release (secrete) hormones.

A **target organ** is the organ that a specific hormone acts upon.

People often link **hormones** with teenage mood swings and puberty. In this topic you will learn about hormones and how they control processes in your body. If something scares you, your heart beats faster. This happens because a chemical called adrenaline is released into your bloodstream. Adrenaline is a hormone. It travels to your brain which responds by sending a nerve impulse to your heart. The nerve impulse makes your heart beat faster. Organs that release hormones are called **glands**. They act on specific organs called **target organs**.

Hormone – Hormones are chemical. They travel in the blood to their
Glands – Glands are organs that are release (sec
Target organ –

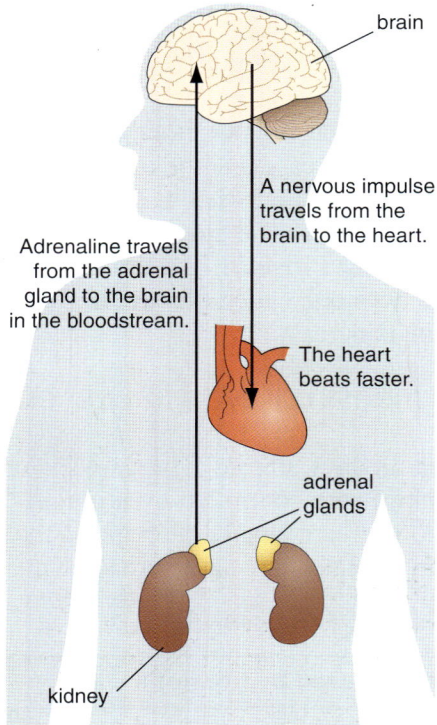

brain

A nervous impulse travels from the brain to the heart.

Adrenaline travels from the adrenal gland to the brain in the bloodstream.

The heart beats faster.

adrenal glands

kidney

Figure 1.12 How the release of adrenaline can affect the heart

⑨ Look at Figure 1.12. It shows how adrenaline speeds up the heart.

Copy and complete this table to name the gland, hormone and target organ for this response.

Gland	adrenal.
Hormone	adrenalin
Target organ	heart.

⑩ How are hormones transported around the body?

How does a hormone control the water content of your body?

Your body needs the right amount of water. You take in water when you eat and drink. Water is lost when you breathe out and when you sweat. Water is also lost in urine when you go to the toilet.

Your kidneys control the amount of water in your body. This response in the kidneys is controlled by anti-diuretic hormone (ADH). ADH is secreted by the pituitary gland. It is carried in the blood to the kidneys, which are its target organs. More ADH causes the kidneys to leave more water in the blood. Less ADH causes the kidneys to remove more water from the blood and release it into the urine (Figure 1.14).

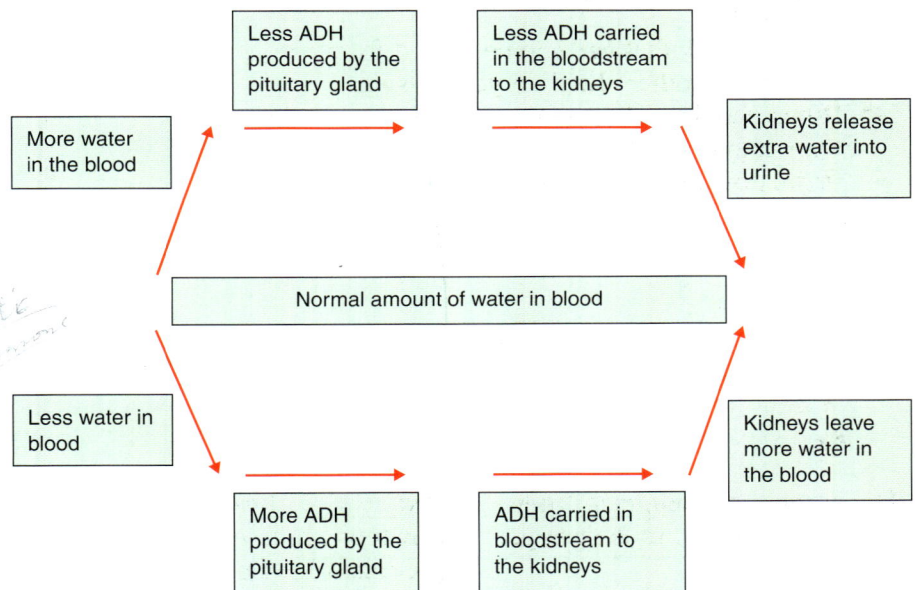

Figure 1.13 A flow chart showing how water in the blood is controlled by the hormone ADH and the kidneys

⑪ Copy and complete the sentences below. On a hot day, water is lost from the body in sweating. Water is also lost in urine when you go to the toilet. When the amount of water in your blood decreases, more anti-diuretic hormone is produced. The ADH is carried in the blood to the kidney. Less water is then put in the _____.

The lining of the womb (uterus) thickens

Hormone levels change at the end of the cycle to prepare the womb lining

An egg matures in an ovary

An egg is released from an ovary

An unfertilised egg is lost with the lining of the womb during a period

A fertilised egg may implant in the lining of the womb

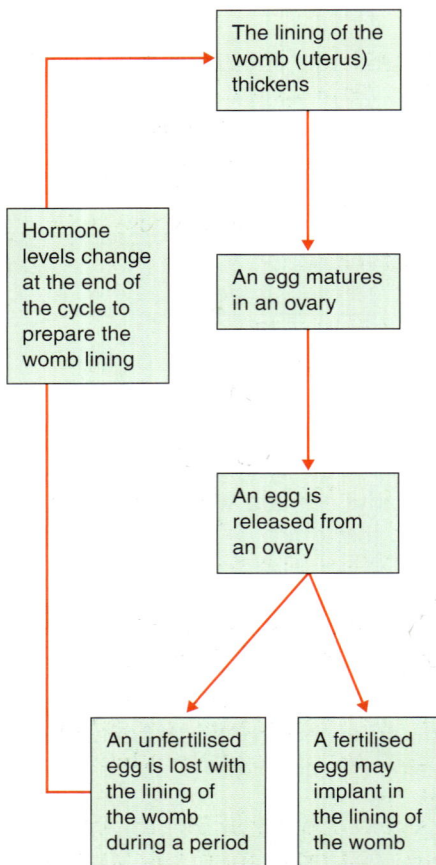

Figure 1.14 A flow chart showing the changes during the menstrual cycle

How do two hormones work together to control the concentration of sugar in your blood?

Your body needs to control the amount of sugar in your blood. When you eat, the amount of sugar in your blood increases. Some of this sugar is stored. It can then be released back into the blood when it is needed. If there is too much sugar it is stored in the liver. The amount of sugar that is stored is controlled by a hormone called **insulin**. The amount of sugar that is released is controlled by a hormone called **glucagon**. Both of these hormones are released from the pancreas.

How do hormones control the menstrual cycle in women?

A woman's menstrual cycle lasts about 28 days. Figure 1.14 shows what happens during the cycle. All these changes are controlled by hormones. The levels of these hormones in the bloodstream change throughout the menstrual cycle. The hormones and the changes they cause are shown in Table 1.2.

Hormone	Released by	Change caused by hormone
Follicle-stimulating hormone (FSH)	Pituitary gland	• Causes an egg to mature in an ovary • Stimulates the ovaries to produce oestrogen
Luteinising hormone (LH)	Pituitary gland	• Stimulates the release of an egg from an ovary
Oestrogen	Ovaries	• Stops further production of FSH • Stimulates the pituitary gland to release LH

Table 1.2 The main hormones that control the menstrual cycle

12 Copy and complete the table below to name the gland, hormones and target organ that control blood sugar level.

Gland	pancreas
Hormones	insulin
Target organ	liver

13 The amount of sugar in a woman's blood (her blood sugar concentration) was measured six times during one day. The following values were obtained, in mg of sugar per 100 cm³ of blood: 92, 89, 95, 97, 115, 97.
a) Copy and complete the following sentences.
   The range of data for the woman's blood sugar concentration varies from ___ (the lowest value) to ___ mg per 100 cm³ of blood (the highest value). One of the values appears to be anomalous. Anomalous values are very different from the others. They do not lie within an expected range or fit an expected pattern. The anomalous value is ___ mg of sugar per 100 cm³ of blood. When average (mean) values of data are calculated, anomalous values are usually ignored. Ignoring the anomalous value, the woman's mean (average) blood sugar concentration is ___ mg per 100 cm³ of blood.
b) State two things that might cause the amount of sugar in your blood to fall.

Activity – Using hormones to control fertility

A woman's fertility tells you how likely she is to become pregnant. If a woman is very fertile, she will conceive easily. The hormones involved in the menstrual cycle can be used to help women who are less fertile to become pregnant. They can also be used to prevent a woman getting pregnant. This is called contraception.

Laura and David have just got married. They plan to have children in about three years' time. Before then they want to have sex without getting pregnant. Their doctor suggested that Laura should go on the pill. The pill contains a hormone which stops an egg maturing in the ovaries each month. The pill is taken every day for 21 days in a 28 day cycle. In the week when Laura is not taking the pill she has a normal period.

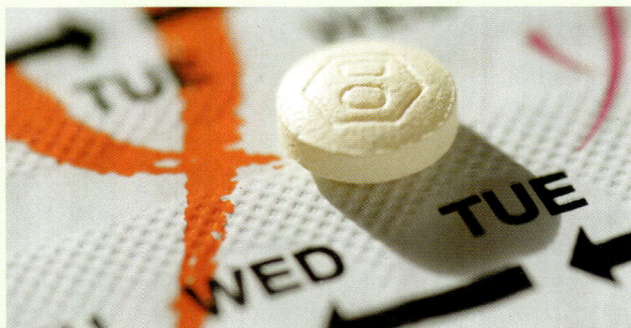

Figure 1.15 The contraceptive pill that Laura takes

❶ Why do you think Laura's doctor suggested the pill as a method of contraception? Discuss this in a pair and write a list of reasons for using the pill.

❷ Go to the website www.fpa.org.uk. Click on 'contraception' and then 'the combined pill' for help with this question. Write down two disadvantages of using the contraceptive pill.

❸ The pill contains the hormone oestrogen. Use Table 1.2 to explain how oestrogen can prevent an egg from maturing.

After three years, Laura and David decided to start a family. Laura stopped taking the pill. They had sex regularly but after a year Laura was still not pregnant. Her doctor referred them to a fertility clinic. After some tests, the doctor found that

Laura was not getting pregnant because eggs were not maturing properly in her ovaries. This was caused by a low concentration of one of her hormones. She was given a fertility drug that contained this hormone. The hormone caused her eggs to mature properly in the ovaries. After six months, Laura became pregnant.

❹ The fertility drug caused Laura's eggs to mature properly. Look at Table 1.2. What hormone do you think the fertility drug contained?

❺ a) Some people do not agree with the use of hormones to treat infertility (being unable to get pregnant). Their reasons may be based on: A evidence, B hearsay, C personal opinion.
Look carefully at the reasons given by the following three people for not agreeing with this use of hormones. In each case, say which of A, B or C their reason is based on.
 i) Kelly says 'It's wrong to interfere with nature.'
 ii) Her husband, Ali, says 'The use of hormones can make for problems in a woman's hormonal system.'
 iii) Becky, their friend, says 'It's against my religious beliefs.'
 b) Write two sentences to explain your own views on the use of hormones to treat infertility.
 c) What do you think we should base our views on?

In some cases of infertility just taking a hormone does not help a couple to have children. Another treatment called IVF (*in vitro* fertilisation) is sometimes used. This involves the woman taking a hormone to increase egg production. Then eggs are removed from her ovaries with a needle. They are fertilised with the man's sperm in a laboratory. Two of the fertilised eggs are then inserted into the woman's womb.

❻ State one cause of infertility that IVF can be used to treat.

❼ Suggest two problems that may result from IVF.

1.3 How do our bodies maintain the right temperature?

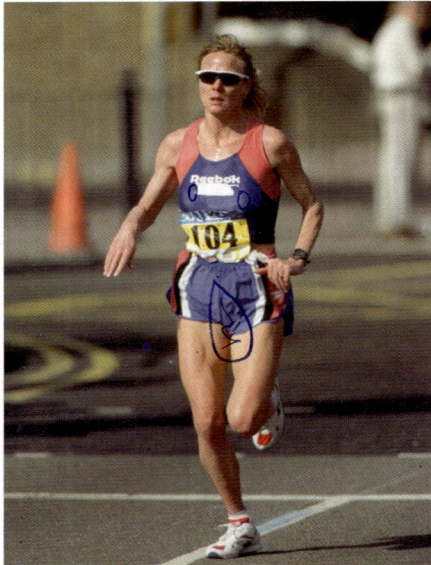

Figure 1.16 A marathon runner on a hot day

Internal body temperature (°C)	Symptoms
28	Muscle failure
30	Loss of body temperature control
33	Loss of consciousness
37	Normal
42	Central nervous system breakdown
44	Death

Table 1.3 Symptoms that occur when the body's internal temperature changes

Table 1.3 shows that body temperature can have very serious effects on how your body works. Fortunately your body has several ways of controlling body temperature at its normal value of 37 °C. It can cool itself down if it gets too hot and warm up if it gets too cold.

How does the body cool down?

Your body warms up on a hot day or when you exercise. Imagine a marathon runner on a hot day.

When you sweat, your body cools down. Sweat is produced by sweat glands in the skin. As the sweat evaporates off your skin, it transfers some heat away from your body. This cools down your body.

Your skin becomes red when you are hot. This happens because more blood flows near the surface of your skin. Some of the heat that the blood carries to the surface is then lost to the air.

15 Look at Figure 1.16. Write down two things that the runner's body is doing to cool down.

16 How do you think these changes reduce her body temperature?

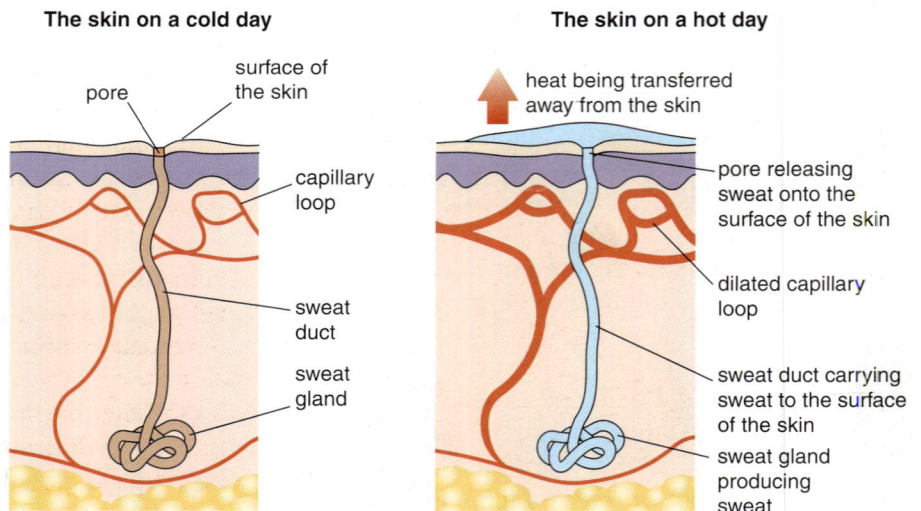

Figure 1.17 The surface of the skin changes when we need to cool down.

How does the body warm up?

On a cold day, your body temperature will start to fall. Your body will respond to this by trying to prevent heat being lost. It stops sweating and reduces the blood flowing near the surface of the skin. Your body also produces more heat by shivering. Shivering causes your muscles to contract and then relax very quickly.

⓱ Copy and complete the table below.

Ways the body warms up	Ways the body cools down

⓲ When the body gets far too cold, the person suffers from hypothermia. This happens when the body temperature falls well below the normal 37 °C. Many more old people than young people die from hypothermia in the winter in their own homes.

Write down two possible reasons for this.

1.4 What are drugs and how do drugs affect the body?

Drugs affect processes in the body. **Medicines** are drugs that are helpful (beneficial) to the body. Some people use substances like nicotine, alcohol, cannabis and heroin as recreational drugs. Some of these are more harmful than others. It is important to know how a drug affects your body in order to understand its risks as well as its benefits.

Medical drugs

Medical drugs or medicines are used to treat diseases, injuries and pain. Medicines are beneficial when they are used properly. They can be harmful if misused. Paracetamol is a good painkiller, but if you take more than the correct dose, it can damage the liver. You should always follow the instructions given with any medicines.

Medicines have been used since prehistoric times. Cave paintings from 8000 BC show tribal healers giving people plants to eat and cure illness. The ancient Egyptians had doctors who used herbal medicines. Many medicines are made from natural substances found in plants. Some of these plants grow in the rainforest. For example, quinine is extracted from the cinchona tree, found in South America (Figure 1.18a)). It is used to treat malaria. A plant called rosy periwinkle from Madagascar provides two of the most important chemicals used to treat cancer. New drugs are still being developed from natural substances found in plants.

Figure 1.18 a) The cinchona tree and b) rosy periwinkle plant are used to produce essential medicines.

Figure 1.19 There are many different medicines available today.

Figure 1.20 A person suffering from the side effects of thalidomide

Today, all new drugs are thoroughly tested before they are used on patients. Testing has four important stages.

Stage 1: Drugs are tested in the laboratory. They are tested on live cells and sometimes on animals.

Stage 2: Drugs that seem to be effective are tested on a small number of healthy humans.

Stage 3: The drug is tested on about 100 humans who suffer from the disease.

Stage 4: Finally, the drug is tested on up to several thousand sufferers. It is often compared with a drug that has already been found to be effective.

Only drugs that pass all safety checks can be launched as a new product. Very occasionally a drug that has been tested is later to found to have harmful side effects. A drug called thalidomide was used in the 1960s as a sleeping pill. It was also used to relieve morning sickness in pregnant women. Sadly, many babies born to mothers who took this drug had very under-developed arms and legs. The drug was then banned. Since then it has been used successfully to treat leprosy.

The testing and development of modern medicines and drugs shows how research scientists work. The process often begins with an idea or an explanation why a particular drug might be used to treat a particular disease. This idea is called a **hypothesis**. The hypothesis can be used to predict the effects of the drug to be tested.

The first tests involve laboratory experiments before any trials on humans.

If the results of the experiments support the hypothesis, then **clinical trials** can begin. These trials use carefully selected groups of people to check the effects of the drug. People of different sex, age, weight and lifestyle are chosen. The tests also involve control groups to ensure that only the effects of the drug (the independent variable) are observed. The control group are given a placebo. This could be a tablet that looks like the drug but has no effect. None of the people in the trials are told whether they are taking the experimental drug or the placebo.

If, after all the tests and trials, the results show that the drug is beneficial to patients, then large-scale production of the drug can begin.

⑲ Why is it important to follow the instructions carefully when you take medicines?

⑳ Write down one advantage of testing a new medicine on live human cells instead of animals.

㉑ Why do scientists compare a new drug with an existing drug for the same disease?

㉒ Write down one side effect of the drug thalidomide.

Recreational drugs

How does smoking tobacco affect the body?

Tobacco smoke contains over 4000 different chemicals. Many of these chemicals can kill you. Nicotine, tar and carbon monoxide are just three of the dangerous chemicals in tobacco smoke. Table 1.4 shows their effects on the body.

Chemical	Effect
Nicotine	Highly addictive
Tar	Causes lung cancer, bronchitis and emphysema
Carbon monoxide	Reduces the amount of oxygen in the blood

Table 1.4 Three chemicals found in tobacco smoke and their effects on the body

Cigarette companies did not have to put health warnings on their packets until 1971.

Activity – What does research tell us about the use of cannabis?

Recently there has been a lot of research into two questions about the use of cannabis.

- Do cannabis users move on to use harder drugs?
- Can smoking cannabis cause psychological (mental) problems?

Read the following quotes from different researchers.
i)   'Cannabis does not lead to experimentation with harder drugs.'
ii) 'Cannabis does not act as a "gateway" drug. Measures to curb cannabis use do not have a knock-on effect on the use of harder drugs.'
iii) 'Many hard drug users have followed the same path from cigarettes and alcohol, to cannabis, to heroin and cocaine.'
iv) 'Early marijuana smokers were found to be about five times more likely to move to harder drugs than their twins who did not smoke marijuana.'
v)   'Some studies have suggested that long-term cannabis use can increase your risk of developing a personality disorder called schizophrenia.'
vi) 'Smoking cannabis doubles the risk of developing mental illnesses.'
vii) 'There was no proof of any link between taking cannabis and mental illness.'

❶ Write down two quotes that say the opposite to each other.
❷ Write down three important points to summarise the quotes.
❸ Researchers do not all agree on the effects of cannabis. In a pair, discuss why you think this is. Write down some reasons.
❹ Newspapers often use quotes from researchers in their headlines.
   Explain how this could mislead the public.
❺ Imagine you are doing research to see if there is a link between cannabis smoking and mental illness.
   a) Who will be involved in the project?
   b) How many people will be involved?
   c) How long should the study last?
   d) How will you collect information from the people involved?
   e) How will you share your findings with other people?

Figure 1.21 Tobacco products must now carry government health warnings.

Here are some key dates relating to smoking and health.

1951: Start of the first large-scale study into smoking and lung cancer.

1954: A link between smoking and lung cancer is proved.

1957: The British Medical Research Council announces a direct link between smoking and lung cancer.

1962: Another report confirms the link between smoking and lung cancer. The report recommends tougher laws on cigarette sales, cigarette advertising and smoking in public places.

1965: The British government bans cigarette advertising on television.

1971: Government health warnings are required on all cigarette packets sold in the UK.

1976: A 20-year study shows that one in three smokers die from the habit.

1983: A report from the Royal College of Physicians states that more than 100 000 people die every year in the UK from smoking.

1988: Research shows that breathing in other people's smoke can increase the risk of lung cancer.

1989: A UK court rules that injury caused by breathing in other people's smoke can be counted as an industrial accident.

24 It was 20 years after the first major study into smoking before health warnings were put on cigarette packets.

Why do you think it took so long?

25 Which two pieces of information would you tell someone to try to persuade them to give up smoking?

26 Find out about ways that people use to help them give up smoking. Write down two ways and explain how each one helps a smoker to give up.

What other drugs may harm the body?

Drug	Classification	Effects on the body
Alcohol	It is illegal for anyone under the age of 18 to buy alcoholic drinks It is illegal for someone over 18 to supply alcoholic drinks to someone under 18	• Slows down reaction times • Feel relaxed • Lose inhibitions • Can become loud and sometimes aggressive • Unconsciousness • Increased blood pressure • Possible liver and brain damage
Solvents (glue, lighter fluid, paint thinners, correcting fluid)	It is illegal for a retailer to sell a solvent to anyone under the age of 18 if they believe it will be used for inhaling to cause intoxication	• Brief feeling of euphoria • Lose inhibitions • Blurred vision • Dizziness • Blackouts • Possible lung, liver and brain damage
Cocaine	Class A or 'hard' drug. Illegal to possess or sell	• Feeling of well-being and confidence • Increased heart rate and blood pressure • Highly addictive • Depression • Mental health problems
Heroin	Class A or 'hard' drug. Illegal to possess or sell.	• Feeling of well-being • Drowsiness • Blurred vision • Vomiting • Highly addictive • Can cause a coma or death when taken in large amounts

Table 1.5 The classification and effects of some drugs

❷❼ Copy and complete the following sentences.

You must be _____ years old to buy alcohol. One short-term effect of alcohol is _____. A long-term effect of alcohol is _____. _____ is highly addictive. Solvent abuse can damage the _____, _____ and _____.

❷❽ Treatment for drug addicts is free on the National Health Service.
a) Write down two points in support of this policy.
b) Write down one point which criticises this policy.

❷❾ Suggest three things that could be done to reduce the number of drug addicts in the UK.

Summary

✓ Our **nervous system** enables us to react to **stimuli** from our surroundings.

✓ The nervous system includes **receptors**, **neurones**, the spinal cord, the brain and **effectors** (muscles and glands).

✓ Receptors detect stimuli. Stimuli include light, sound, changes in position, chemicals, touch, pressure, pain and temperature.

✓ Information from receptors passes along neurones to the brain. The information travels as electrical impulses.

✓ **Reflex actions** are automatic. They take place very quickly. They involve sensory, relay and motor neurones.

✓ Chemicals called **hormones** help to control conditions inside the body. These conditions include the water content of the body, body temperature and blood sugar levels.

✓ Hormones are secreted by glands. They are transported by the bloodstream to their **target organ**.

✓ Several hormones are involved in controlling the menstrual cycle. These hormones include FSH, oestrogen and LH.

✓ Hormones can be used to control **fertility**. Hormones are used in the contraceptive pill to stop a woman's eggs maturing. FSH is used as a fertility drug to stimulate eggs to mature.

✓ **Drugs** are chemicals that have an effect on processes in the body.

✓ **Medicines** are drugs that are beneficial to people. Other drugs can harm the body.

✓ Many drugs are made from natural substances. When new drugs are developed they must be thoroughly tested.

✓ New drugs are first tested in the laboratory and then on human volunteers.

✓ Occasionally drugs that have been tested cause unexpected side effects. Thalidomide was used to stop 'morning sickness' in pregnant women. Unfortunately, thalidomide caused babies to have deformed arms and legs and so it was no longer given to pregnant women.

✓ Heroin, cocaine and nicotine are examples of addictive drugs.

✓ Claims have been made about the effects of cannabis on health. Cannabis has been linked to taking hard drugs. However, research has led to different conclusions about these claims.

✓ The link between smoking tobacco and lung cancer took many years before it was accepted.

✓ Tobacco smoke contains a number of chemicals that damage health.

✓ Alcohol affects the nervous system. It slows down reactions and helps people to relax. But too much alcohol can lead to a lack of self-control and even unconsciousness. Alcohol can also cause liver and brain damage.

EXAMQUESTIONS

1 A student accidentally touches a hot saucepan. Her hand moves automatically away from the pan.

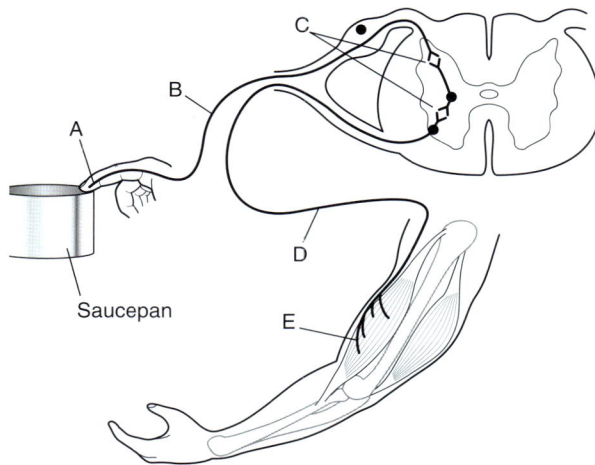

Figure 1.22

a) In this reflex action:
   i) where is the receptor?
   ii) where is the effector? *(2 marks)*
b) Explain how an impulse crosses the synapse labelled C. *(1 mark)*
c) Explain why this type of reflex action is important to our bodies. *(1 mark)*

2 A hormone is involved in controlling the water content of the body.
a) State **two** ways in which water leaves the body. *(2 marks)*
b) What can cause a rapid fall in the water content of the body? *(2 marks)*
c) Explain how the body responds to a fall in water content. *(3 marks)*

3 Read the information about contraceptive implants.

A contraceptive implant works a bit like a contraceptive pill. It contains a hormone present in contraceptive pills, which prevents pregnancy. The hormone implant is a small, thin flexible rod, 4 cm long and made of plastic. It is inserted just under the skin of a woman's arm. This procedure must always be undertaken by a doctor who is familiar with the

technique. The implant gradually releases a small amount of the hormone, which prevents pregnancy for up to three years.
a) How can a hormone prevent pregnancy. *(2 marks)*
b) Give **one** drawback of using contraceptive implants rather than contraceptive pills. *(1 mark)*
c) Contraceptive implants are being used increasingly in birth control programmes in developing countries, instead of contraceptive pills.
   Can you think of a reason for this? *(2 marks)*

4 A person's blood sugar level rises after a meal. Following this the pancreas secretes a hormone called insulin. This causes the liver to store sugar until it is needed.
a) i) Name the gland involved in this response. *(1 mark)*
   ii) Name the target organ involved in this response. *(1 mark)*
b) Explain how insulin is transported from the pancreas to the liver. *(2 marks)*
c) Why does the liver need to release sugar when a person is taking vigorous exercise? *(1 mark)*

5 Medicines must be thoroughly tested before they can be prescribed to the public.
a) Outline the procedure for testing a new medicine. *(2 marks)*
b) A drug called thalidomide caused severe side effects despite passing safety tests and clinical trials.
   i) Describe the side effects caused by thalidomide. *(2 marks)*
   ii) Explain why thorough testing did not identify the possibility of these side effects. *(3 marks)*

6 Alcohol and tobacco can damage the body.
a) Give one example of how the body is damaged by tobacco smoke. *(1 mark)*
b) Why do you think that people find it difficult to give up smoking? *(2 marks)*
c) Why should motorists not drive after they have been drinking alcohol? *(2 marks)*

Chapter 2
What can we do to keep healthy?

At the end of this chapter you should:

✓ understand the importance of different food groups for your health;

✓ be able to discuss the different types of fat and their effect on cholesterol levels;

✓ understand how the needs for energy and nutrients change with age;

✓ understand the link between exercise, food energy and health;

✓ know how to calculate the amount of energy used during exercise;

✓ be able to decide whether a diet is healthy;

✓ be able to explain the defences we have against micro-organisms which cause disease;

✓ understand the importance of immunisation programmes;

✓ have considered the value of antibiotics and their problems.

Figure 2.1 Food can be colourful and interesting.

2.1 Eating for health

A varied diet

Do you eat the same foods every day? Do you think about your choice of food?

Most people eat a varied diet that includes everything that is needed to keep their body healthy. The important word in the last sentence is 'varied' because no single food can provide all the essential nutrients your body needs to work well. An easy way to make sure your diet contains all the essential vitamins and minerals is to eat as many different naturally coloured foods as possible. Red tomatoes, green broccoli and purple plums are three examples.

For a healthy diet you should try to eat some food from each group in Table 2.1 every day.

Amino acids are molecules that build up into proteins.

Fatty acids are part of fat molecules.

1 a) List six different-coloured fruits or vegetables. Try to include all the colours of the rainbow.
b) Name eight of the fruits and vegetables shown in Figure 2.1.

Food group		Examples	Health points
Energy foods	Figure 2.2	• Whole grains such as rice, millet • Yams, potatoes • Cereals, bread and pasta	• Whole grains provide extra fibre. In high-fibre foods, starches are broken down slowly and this releases sugars over a longer period. This helps to prevent snacking.
Food for building bones and teeth (Two or three servings daily)	Figure 2.3	• Dairy products such as milk, cheese, yoghurt, fromage frais • Soya or rice milk with added calcium	• Your skeleton increases in strength most rapidly during your teens. • Skimmed and semi-skimmed milk contain as much calcium as other milk but much less fat.
Muscle-building foods	Figure 2.4	• Red meat, poultry, sausages, bacon, eggs • Nuts, beans, lentils • Fish such as salmon, sardines, cod	• Vegetarians must eat vegetable proteins from several different foods to get all the essential **amino acids**. Red meat eaters should avoid too much fat. • Oily fish such as sardines and salmon provide the essential **fatty acids** which are good for the brain. • Meat and fish contain vitamins A, D and E.
Foods for clear skin and a healthy digestive system (A minimum of two portions of fruit and three portions of vegetables daily. One portion is about 80 g.)	Figure 2.5	• Leafy vegetables such as cabbage, broccoli • Peas, beans, courgettes, squash, onions • Salad such as lettuce, cucumber, peppers, tomatoes • Fruit such as oranges, nectarines, strawberries, mangos, blue-berries	• The coloured pigments in fruit and vegetables contain minerals and vitamins. Fruits are rich in vitamin C. • The fibre in fruit and vegetables 1 helps the food to move quickly through the intestine and prevents colon cancer 2 slows down the release of sugars – this reduces the risk of diabetes and prevents snacking

Table 2.1 The health points of different food groups

If you don't eat all the essential nutrients you will become **malnourished**, no matter what weight you are. The foods listed in Table 2.1 will give you all the vitamins and minerals you need. But your meals must also give you enough energy.

Energy values of different foods are given in joules (J) or kilojoules (kJ). The energy values are written on the 'nutritional information' panels on food packets. In some cases, the old units, calories or kilocalories, are still used.

2 If a child refused to eat fruit and vegetables for a few weeks, what health problems might he or she develop?

How do we compare the energy contents of food?

The experiment described below will help you to understand how manufacturers work out the energy values of different foods.

To find the energy values of different foods in a simple laboratory experiment, you could take pieces of each food, set fire to them, hold the burning food under a test-tube of water, and see which food heats the water most (Figure 2.7).

Before starting an experiment such as this, you need to ask yourself some important questions. Is the experiment accurate or am I likely to make lots of errors? Is the experiment fair? Am I treating each food in the same way?

To make this experiment a **fair test** you should:
- make sure the type of food is the only variable;
- use the same mass of food for each test;
- use the same volume of water for each test;
- stir the water to spread the heat;
- put screens around the apparatus to stop draughts blowing the flame.

In addition to these points, there are three other problems to think about.
- It can take quite a lot of heating to start the food burning in air. If you heated the foods with a Bunsen burner for different lengths of time to get them burning, this would be an additional variable.
- Not all the heat from the burning food goes into the water. Some will heat the glass of the test-tube and some will go into the air. This will cause **systematic errors**.
- Glass is not a good conductor. It might be better to use a metal beaker to improve heat transfer.

Figure 2.6 A little of what you fancy does you good. Too much makes you fat!

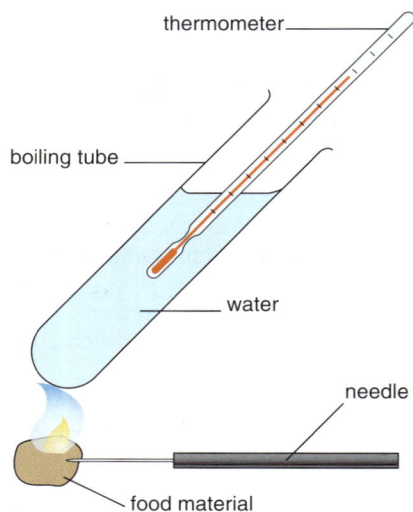
Figure 2.7 Simple apparatus to find the energy value of a food

> In a **fair test** only the variable being tested is allowed to affect the result. All the other variables have been kept constant.
>
> **Systematic errors** affect all results in an experiment. The results are shifted away from the true value because the apparatus has an error.

Now look at Figure 2.8. This shows a diagram of a food calorimeter, used by scientists to find the energy values of foods. Notice how it solves many of the problems with your simple apparatus.

- The food is set alight with a hot wire.
- The food is burned in oxygen, not air, so it will burn more easily.
- The food is burned in a metal container. This will conduct the heat to the water quickly.
- The water container is surrounded by insulation. Very little heat can escape from the apparatus.
- There is a stirrer to spread the heat around the water.

Which apparatus will give the more accurate results?

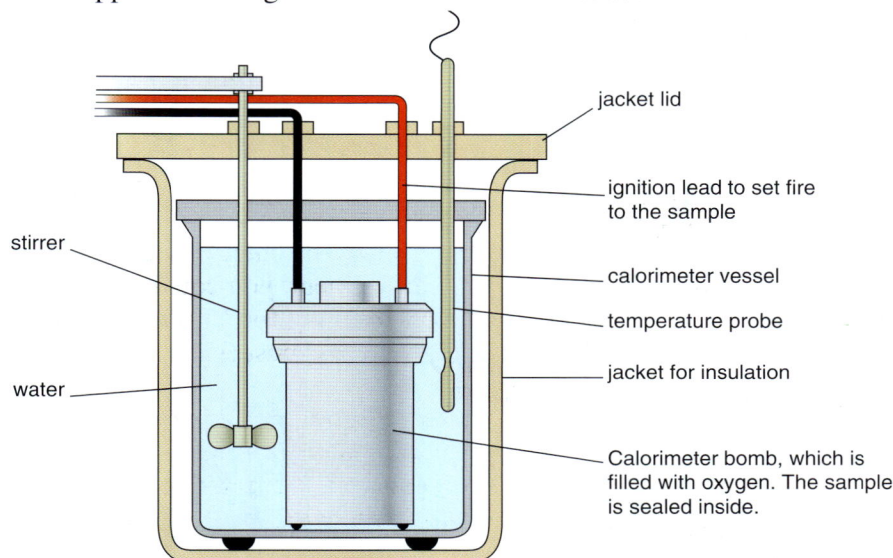

jacket lid

ignition lead to set fire to the sample

stirrer

calorimeter vessel

temperature probe

water

jacket for insulation

Calorimeter bomb, which is filled with oxygen. The sample is sealed inside.

Figure 2.8 A food calorimeter used by food scientists to find the energy values of foods

Experiments with food calorimeters show that one gram of fat contains more than twice as much energy as one gram of carbohydrate.

Is fat bad for us?

The answer to this question is 'no', because fat is good for us. But it is bad for us when we eat too much of it. When someone eats far too much fat, he or she becomes seriously overweight (obese).

It is important to eat fats because:
- they protect vital organs (such as the kidneys) by cushioning them from damage;
- they form an energy reserve;
- essential fatty acids (EFAs) are used in our bodies to make some hormones, including the sex hormones;
- cholesterol is important for making strong cell membranes;
- they contain the fat-soluble vitamins (A, D, E), which are important for good health.

On the side of food packets you will find a 'nutritional information' panel. This may have a list of the different types of fat in the food.

Sorting out the fats

You should understand the terms saturated fats, mono-unsaturated fats and polyunsaturated fats.

Saturated fats are hard at room temperature. They include the fat around meat or fat left in the cold pan after cooking.

All unsaturated fats will help to reduce cholesterol. Polyunsaturated fats contain the EFAs. Figures 2.9, 2.10 and 2.11 will help you sort out the 'good' and 'not so good' fats. Saturated and unsaturated fats are studied more fully in Chapter 7.

Figure 2.9 Dairy products and red meats contain saturated fats. Saturated fats can be used for energy. Eating too much saturated fat makes you fat and can raise your blood cholesterol level.

Figure 2.10 Olive oil, peanuts and avocados contain mono-unsaturated fats. These can be used for energy and help to lower blood cholesterol levels.

Figure 2.11 Oily fish (such as salmon and sardines) and vegetable oils contain polyunsaturated fats. They also contain essential fatty acids. Eating oily fish helps to lower the blood cholesterol level.

Cholesterol is important in the body for making strong cell membranes and vital hormones. The cholesterol level in the blood depends on:
- the amount produced by the liver, and
- the amount eaten in food.

If the cholesterol level in the blood is too high, fat can be deposited in the lining of blood vessels. This reduces the inside diameter of the blood vessels, so it raises blood pressure and increases the risk of heart disease. Men are more at risk of blocked blood vessels than women, especially if they eat too much saturated fat.

❸ Nutritionists say that no more than 10% of the daily energy intake for an active teenager should be saturated fats. Why do they suggest such a low intake of saturated fat?

❹ Nutritionists also suggest that we should eat oily fish, such as salmon or sardines, at least twice a week. Why do they give this advice?

5 Although more and more people are obese, the amount of energy in the food that people eat has decreased slightly. What changes in our lifestyle could have caused more people in rich countries to become obese?

6 What sorts of food contain cholesterol?

7 Some diets are advertised as 'cholesterol free'.
   a) What important role does cholesterol have in cells?
   b) Why is a cholesterol-free diet unhealthy?

8 a) Give five reasons why your diet should include some fats.
   b) Which foods that contain fats do you think you should eat?

Figure 2.12 Save your heart! Don't add salt to food at the table or in cooking. Be more adventurous – add herbs for flavouring.

Using the nutritional information on food packets

Chocolate contains 30 g of fat in every 100 g of chocolate.

A small bar of chocolate has a mass of 50 g, so the total mass of fat in the chocolate bar is 15 g.

(If saturated fats are not listed separately, assume they make up half of the total fats.)

Therefore the mass of saturated fats in the chocolate bar is $\frac{15}{2} = 7.5$ g.

1 g of fat provides 37 kJ of energy.

Therefore the saturated fats in the chocolate bar provide $7.5 \times 37 = 277.5$ kJ of energy.

An average teenage girl needs 8100 kJ energy per day. 10% of this energy should be saturated fats – this is $\frac{8100}{10} = 810$ kJ.

The small chocolate bar provides 277.5 kJ out of the total 810 kJ recommended daily intake of saturated fat.

So the small chocolate bar provides $\frac{277.5}{810} \times 100 = 34\%$ of the recommended daily intake of saturated fat.

Just think of the fat intake if she had already eaten a bag of chips and was planning to have a thick milk shake as well!

Salt in our diet

Salt is added to food to preserve it and to 'bring out the flavour'. It is very easy to eat too much salt. The recommended daily salt intake is 5 g for a 7–14 year old and 6 g for those aged 15 and over. If children eat lots of crisps they will have too much salt. 'Ready-to-eat' meals also have too much salt. There is very little salt in raw and fresh foods.

Doctors are worried about the amount of salt that people have in their diet. Too much salt is linked to high blood pressure. This can lead to heart disease and stroke. Heart disease is a major cause of death in the UK.

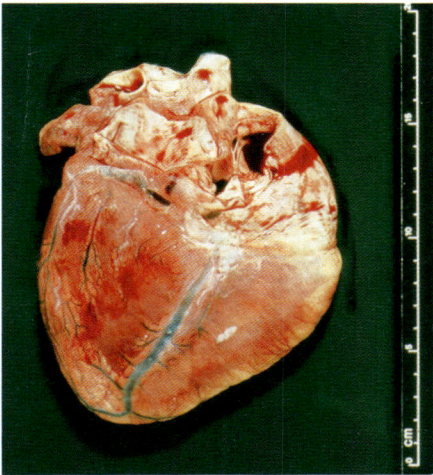

Activity – What is the proportion of fat in your diet?

❶ Keep a food diary for one day. Write down what you have eaten immediately after each meal and snack, before you forget.

❷ Look at the 'nutritional information' labels on the packaging of the foods you eat. Find the amount and type of fat in your food. Make a table like the one below. You will need one row for each different food eaten.

Mass of food eaten (g)	Type of fat (saturated or unsaturated)	Mass of all fat (g)	Mass of saturated fat (g)
Reduced fat crisps 25 g	Both	6.2	2.9

Table 2.2 My food diary

Remember: if the saturated fats are not listed separately, assume they are half of the total fats.

❸ Add up the third and fourth columns in your table to find the total mass of all fats and the total mass of saturated fats.

❹ Assume that your recommended daily energy intake is 9000 kJ. Is your total saturated fat intake more than 10% of this?

❺ Which food gives you most saturated fat?

❻ What problems might you have if your total fat intake is too low?

❼ What foods should you eat in limited amounts if you want to 'eat for health'?

❽ Suggest five ways to reduce your fat intake.

Energy and nutrient requirements change with age

The food you eat is changed into energy. The more active you are, the more food energy you need.

Figure 2.13 Pre-school children are small but growing and very active. They need a varied diet which provides:
- enough energy for running and playing;
- a good supply of protein for growth;
- calcium for making bones and teeth;
- iron for haemoglobin in the blood;
- the fat-soluble vitamins A and D for healthy eyes and bones.

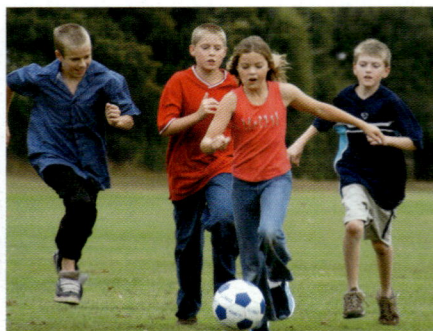

Figure 2.14 Pre-teens should:
- eat foods of all types;
- eat more starchy foods for energy than pre-school children (more wholegrain cereals and potatoes);
- avoid too much fatty food such as cakes, biscuits, chocolate and ice-cream.

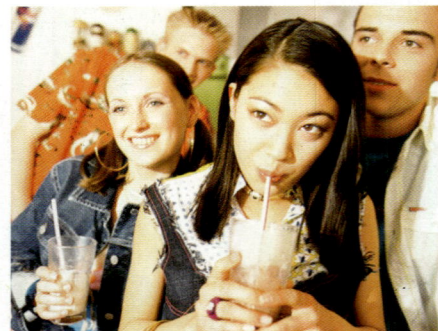

Figure 2.15 Teenagers should think about what they are eating. The teenage growth spurt needs to be supported by eating:
- more protein foods;
- more iron for muscle building in boys and menstruation (periods) in girls;
- enough calcium-rich foods to help bones develop;
- 'five-a-day fruit and vegetables' to provide vitamins and minerals for general good health and a clear skin.

Figure 2.16 **Pregnant and breast-feeding mothers** are eating for themselves and a baby (not for two adults). They will therefore need:
- more calcium for bones and teeth;
- a little more protein;
- more fruit and vegetables to provide plenty of vitamins and minerals.

Figure 2.17 **Adults** need to eat a balanced diet and avoid excess fat, sugar and alcohol. Their energy intake should depend on how much exercise they get. For example, a window cleaner will need to eat more than an office worker.

Figure 2.18 **Older people over 65** tend to eat a less varied diet and often don't get enough vitamins and minerals. They should:
- continue to eat a range of fruit and vegetables;
- eat some protein every day;
- change their energy intake to balance the exercise they take.

⑨ More and more children are becoming obese. This will cause serious health problems when they get older. Copy and complete the table below showing which of the lifestyle choices will or will not increase the risk of obesity if done several times each week.

Lifestyle choice	Will this increase the risk of obesity? Yes or no
Skateboarding	No
Eating chips while watching TV	
Playing computer games for hours	
Playing active games outside	
Reading all evening	
'Eat all you want' pizza evenings	
Playing regular sport	
Swimming	
Playing the piano for hours	

⑩ Which foods should we eat to reduce the risk of colon or bowel cancer?

⑪ a) Which foods should a pregnant woman eat more of?
   b) Write down a reason for eating more of each of these foods.

⑫ a) If you eat several bars of chocolate every week, what might you be eating too much of?
   b) What health problems might result if you regularly ate too many chocolate bars?

⑬ Heart disease is a problem in the UK. What could you do to reduce your risk of heart disease?

⑭ Haemoglobin in red blood cells transports oxygen. Which mineral is needed for the production of haemoglobin?

2.2 Exercise for health

Activity	Energy used in joules (J)
Walking slowly	314
Walking quickly	628
Pushing electric/petrol lawnmower	700
Cycling at a steady speed	750
Aerobics	810
Swimming (steady crawl)	820
Tennis doubles	630
Tennis singles	1000
Running 1 km in 6.2 minutes	1250
Running 1 km in 4.7 minutes	1700

Table 2.3 The energy requirements of different activities. (The table shows the energy a person weighing 60 kg (9 stone 8 lb) would use in 30 minutes of each activity.)

⑮ Look at Table 2.3. How does the speed of an activity affect the amount of energy used?

More energy is used if:
- your weight is greater;
- you are fit with well-developed muscles.

⑯ a) Copy the table below. Work out the total energy used by Helen in a typical day. Assume Helen weighs 60 kg and refer to Table 2.3 for the amounts of energy used by each activity.

Activity	Time (min)	(Show your working here)	Energy used (J)
Helen cycles to school and back home, taking 15 min each way	30		750
On arrival she does one hour of swimming training	60	Energy used in swimming for 30 min is 820 J Energy used in swimming for 60 min 820 × 2 =	
At lunchtime she walks around slowly while chatting	30		
In the evening she plays tennis (doubles) for an hour			
		Total	

Table 2.4 Helen's activity record

b) Copy the table below and calculate the total energy used by Saroj in a typical day. Assume that Saroj weighs 60 kg and refer to Table 2.3 for the amounts of energy used by each activity.

Activity	Time (min)	(Show your working here)	Energy used (J)
The day starts with Saroj's paper round. This takes 30 mins, walking quickly.			
He is late, so he runs the 2.0 km to school, getting there in 15 mins			
He rests at lunchtime		0	0
He walks home slowly, taking an hour to get there			
In the evening he earns more pocket money by mowing an elderly person's lawn. He takes 30 mins using an electric push mower			
		Total	

Table 2.5 Saroj's activity record

c) Now make a similar table to work out your own energy use on a typical school day.

> **Metabolic rate** is the rate at which food is broken down chemically in our bodies. It can be measured in joules per minute.
>
> Muscle tissue has a higher metabolic rate than fatty tissue. That means that 1 kg of muscle uses up more energy than 1 kg of body fat. So a muscular person uses up more energy than a skinny person or a fat person.

Getting the energy balance right

Cars which are fuel-efficient can go a long distance on a litre of petrol. In a similar way, our bodies are made to be food-efficient. Our ancestors were active hunter–gatherers and farmers. They had to search for food and sometimes it was in short supply. Most of the time, they had a good **energy in / energy used** balance. This means the food they ate (**energy in**) balanced the **energy used** in their active daily lives.

In many parts of the world our lifestyle has changed greatly since the time when our ancestors had to find their own food. Today, we usually go to the supermarket in a car and this does not use up much food energy.

- It is rare for us to walk anywhere. We jump into a bus, train or car to get to places.
- TV and cinema adverts urge us to eat high-energy foods such as crisps and canned drinks.
- Many people have office jobs rather than hard physical jobs such as mining and farm labouring.
- We spend lots of time playing computer games, watching TV and DVDs. 'Couch potatoes' and 'mouse potatoes' don't use much energy!

Compared with our ancestors, we have a double disadvantage – **more energy in** (food) and **less energy used** (exercise) in our daily lives. So, we can easily put on weight.

How can we achieve the **energy in = energy used** balance to avoid putting on weight?

To keep the **energy in = energy used** balance and avoid putting on weight, you should:
- take regular exercise and walk more – 4000 steps a day should keep your weight the same;

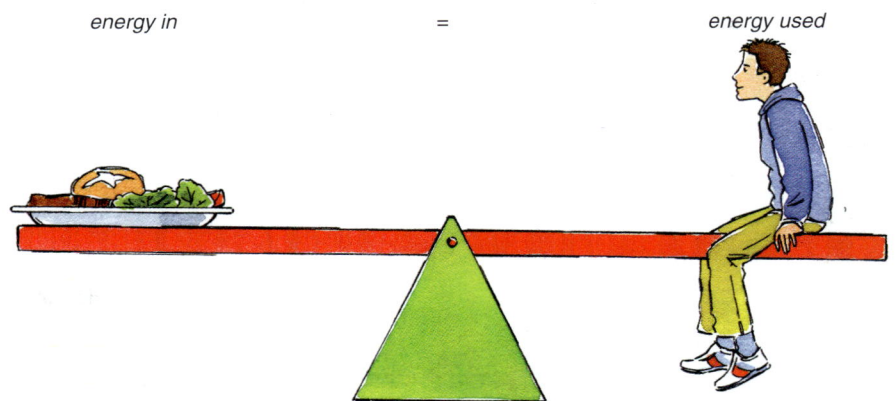

energy in   =   *energy used*

Figure 2.19 To avoid putting on weight, our energy intake must equal the energy we use. Energy in = energy used.

Activity – Food, health and exercise

Work in a group.

① During a typical day, list
   a) the high-energy foods you ate or avoided;
   b) the activity opportunities you took or avoided.

② Is your life healthy or could you make some changes to improve your health? (Remember, any changes must be fun if you are going to keep them up.)

③ Doctors say that small changes such as walking up stairs, taking a slightly longer route to walk home or using a bike instead of going by car are more beneficial than trying to jog 5 miles three times a week. Suggest three reasons for this.

- be aware that some cooking methods (such as frying) increase the energy in foods;
- know which foods contain hidden fats and sugars.

The benefits of exercise

Most gyms have a sudden increase in members around 1 January! Many people are keen to start but then they soon stop attending regularly. Ten good reasons for exercise are given below but remember that exercise needs to be both regular and fun. You won't develop a 'six-pack' overnight, but firm well-toned muscles are well worth the effort.

Exercise is good for you because it:
- increases the strength of your heart muscles (more blood is pumped per beat) and lowers blood pressure;
- improves the strength of your lungs (more air is moved in and out);
- brings more oxygen into the blood, which benefits your brain and your body;
- makes your muscles bigger and more efficient;
- strengthens your ligaments and tendons and reduces the risk of injury;
- makes your bones stronger and reduces the risk of weak bones;
- lowers cholesterol levels in the blood and reduces the risk of blocked blood vessels;
- makes you more resistant to infection;
- raises your metabolic rate and continues 'to burn up food' after you have finished exercising;
- leaves you with a 'feel-good factor'.

If you check back, you can see that exercise and diet can work together to improve the health of your body.

2.3 What happens when the energy balance is upset?

Figure 2.20 'Energy in' is greater than 'energy used'. If our energy intake is greater than the energy we use, we become overweight.

Obesity is a global health issue: 310 million people are affected.

Figure 2.21 'I don't enjoy football, I can't run as fast as the others.'

The calculations in question 16 on page 26 show a clear link between exercise taken and energy used. The key facts are:
- the less exercise you take, the less food you need;
- if you take in more food (energy) than you use, the excess is stored as fat.

In recent years, there has been a lot of publicity about health problems caused by being overweight. At one time, obesity was mainly a problem of 'old age', but now many children and teenagers are overweight.

Being overweight means having a body mass index (BMI) greater than 25. Obesity is defined as having a BMI of 30 or more.

$$\text{Body mass index (BMI)} = \frac{\text{body mass (kg)}}{(\text{height})^2 \ (\text{m}^2)}$$

The BMI is only a rough guide:
- It only applies to people over 18 years old.
- It does not take into account the fact that muscle is denser than fat, so athletes and body builders may have a BMI above 25 and still be healthy.
- It does not account for variations in bone mass.

What are the problems of being overweight?

Overweight people are much more likely than thinner people to suffer from arthritis, diabetes, high blood pressure and heart disease. These were once considered diseases of the over 50s, but since the 1990s these illnesses have been increasing. Diabetes and high blood pressure are even seen in overweight teenagers. Children today consume the same food energy as children 10 years ago, but they take much less exercise. Therefore their energy input and energy use are not balanced.

Diabetes

The number of children with diabetes is increasing. Most of these children (95%) are overweight. Most of them are not having as much exercise as they should. Such a high percentage shows that there is a clear link between
- diabetes and being overweight;
- diabetes and not doing enough exercise.

Because of these links, doctors ask young people with diabetes to stick to a planned diet and take lots of exercise.

Doctors are worried about diabetes in young people because it can lead to high blood pressure, heart disease and damage to the nerves, the eyes and the kidneys.

Being overweight puts extra pressure on the joints. This results in wear of the knee and hip joints. Arthritis sets in and can become a painful problem well before middle age.

What are the problems of being underweight?

If 'energy in' is less than the 'energy needed' for living, this can also cause problems.

Figure 2.22 'Energy in' is less than 'energy used'. If our energy intake is less than the energy we need, over a long time, then we become underweight.

You have probably heard of anorexia. This arises when someone reduces their food intake well below what they need to be healthy. Anorexia affects about 1% of girls and women. It is the third most common long-term illness in teenagers. It has a high death rate.

People also become underweight when there are severe food shortages. This often happens when drought, famine or flooding hit developing countries.

- Underweight people suffer from malnutrition because they are short of protein. Children are affected more than adults. You will have seen on TV starving children with swollen bellies and very thin arms and legs.
- People who are underweight also suffer deficiency diseases because of shortages of vitamins, minerals and essential amino and fatty acids (see Section 2.1).
- Underweight people have a low resistance to infection because they cannot form antibodies (see Section 2.6). Diseases such as measles and cholera spread rapidly through refugee camps.

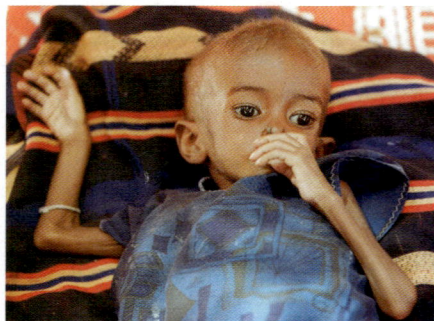

Figure 2.23 This toddler is suffering from a shortage of protein. The child has no fat beneath the skin, is suffering from wasting of the muscles, and is weak.

2.4 Slimming

Anyone who is slimming tries to reduce their energy intake so that it is less than the energy they use. Their aim is to lose weight and look more attractive. The best way to achieve this is to eat wisely and to exercise more. Slimming has become a multi-million pound business. Slimming diets have now been replaced by healthy eating and fitness plans.

Any diet should provide a wide variety of foods with all the nutrients needed for good health. All the food groups in Table 2.1 should be included, particularly fruit and vegetables. It is not a good idea to keep to a diet based on one food, such as bananas or eggs. The long-term aim must be to achieve a healthy eating and exercise plan so that weight remains constant with a BMI between 20 and 24.

In the last four sections, we have seen that what we eat and how active we are can affect our health. In the next four sections we will look at the topic of disease.

Activity – Checking slimming diets

Work in a group.

a) Collect three different slimming diets from magazines or books.
b) Look at each diet carefully and answer the questions in the table below. Copy and complete the table by putting a tick for yes and a cross for no. You will also need to refer to Table 2.1 on page 19.

Diet question	Diet 1	Diet 2	Diet 3
Will all the food groups listed in Table 2.1 be eaten each day?			
Is there the correct proportion of saturated fats?			
Will the essential fatty acids be eaten?			
Are dairy foods included to provide calcium for bones?			
Does the diet include animal protein? (If it's a vegetarian diet, is there a variety of proteins such as nuts, pulses, seeds and dairy foods?)			
Will the diet provide five servings of fruit or vegetables each day?			
Are the foods colourful and 'fun' to eat or are the suggested meals boring and the same each day?			
Will the foods be satisfying and prevent feelings of hunger?			
Is the quantity of salt less than 6 g per day?			
Does the diet also include exercise?			

c) Which diet would you recommend for a teenager who wanted to become healthier? Be ready to explain your choice of diet to the rest of the class.

2.5

What causes disease?

Diseases are caused by bacteria and viruses. These are far too small to be seen with our eyes. They are called microorganisms because they can only be seen with a microscope. Microorganisms which cause diseases are called **pathogens**. Different pathogens affect different parts of our bodies and this helps doctors to identify which disease we have caught.

Viruses

Viruses are the smallest microorganisms. They:
- are taken into cells in the body;
- use the cell contents to **replicate** (form thousands of identical copies);
- damage the cell as they burst out and
- infect nearby cells and repeat the process.

If you get a virus infection of the cells in your throat, the virus replicates in these cells. When the throat cells burst, thousands of virus particles escape to invade nearby throat cells giving you a very sore throat. After a few days your body's immune system will start to destroy the virus and then you will start to feel better.

Bacteria

Bacteria are much larger than viruses, but we still need a microscope to see them. Pathogenic bacteria live in parts of the body such as the nose and throat but not inside cells. Some bacteria produce waste products which act as toxins (poisons). These toxins irritate membranes in the nose and throat causing a sore throat or runny nose.

The importance of hygiene

When we are very young we learn to wash our hands before eating or cleaning a cut. But, it hasn't always been so. In the 1800s many women died of 'blood poisoning' when giving birth.

Ignaz Semmelweiss was a doctor at a poor-mothers Maternity Hospital in Vienna. He was upset by the high number of women who died after childbirth. He watched carefully and made the following observations.
- 29% of mothers died in the ward that doctors ran, but only 3% of mothers died in the ward run by midwives.
- Doctors did not wash their hands between patients. They did not wash their hands after handling mothers who had died in childbirth.
- A doctor who cut his hand by mistake when examining a mother who died of blood poisoning also died of blood poisoning.
- Few mothers died when giving birth at home.
- Few mothers died if they went into hospital after their baby had been born.

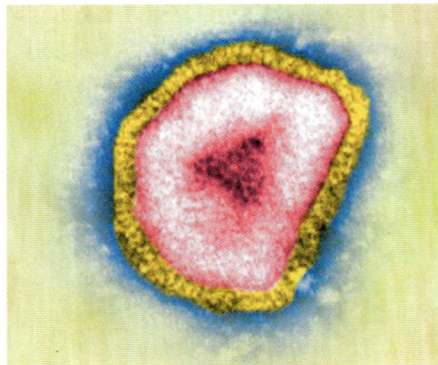

Figure 2.24 Photograph of a virus taken using an electron microscope

Pathogens are disease-causing microorganisms, such as bacteria and viruses.

Replicate means 'form an exact copy'.

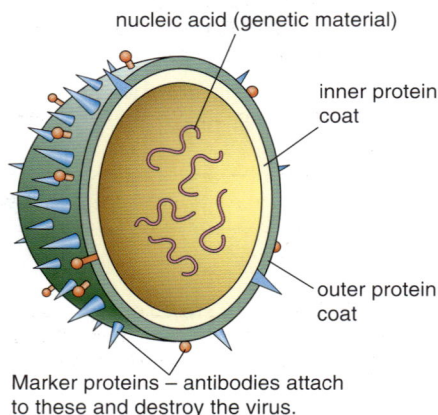

nucleic acid (genetic material)

inner protein coat

outer protein coat

Marker proteins – antibodies attach to these and destroy the virus.

Figure 2.25 The structure of a virus

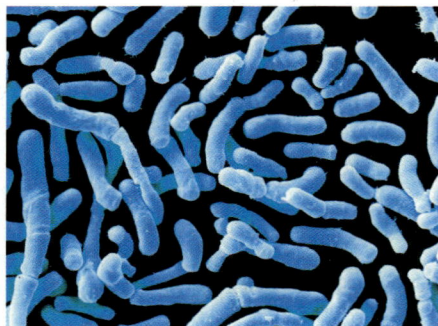

Figure 2.26 A photograph of bacteria taken using an electron microscope

These observations made Semmelweiss think very hard about what caused the deaths of the mothers.

> ⑰ What do you think caused the difference in the death rate in the two wards?

Semmelweiss ordered all doctors to wash their hands between patients. The doctors were annoyed, but when the death rate dropped to 1%, everyone realised he was right. In 1861 Semmelweiss published his findings. They were supported by the data he had collected. Many well-known doctors and scientists laughed at his ideas and he died in 1865 before his simple lifesaving advice of hand-washing was accepted. (See also the information on MRSA in Section 2.7.)

Figure 2.27 Dr Ignaz Philipp Semmelweiss of the Maternity Hospital in Vienna

2.6 The body's natural defence against pathogens

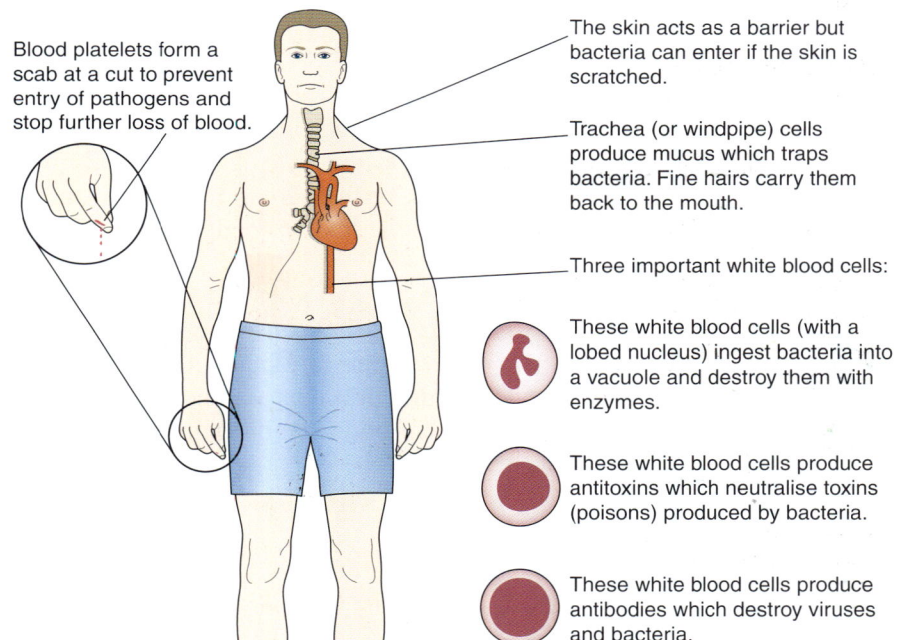

Blood platelets form a scab at a cut to prevent entry of pathogens and stop further loss of blood.

The skin acts as a barrier but bacteria can enter if the skin is scratched.

Trachea (or windpipe) cells produce mucus which traps bacteria. Fine hairs carry them back to the mouth.

Three important white blood cells:

These white blood cells (with a lobed nucleus) ingest bacteria into a vacuole and destroy them with enzymes.

These white blood cells produce antitoxins which neutralise toxins (poisons) produced by bacteria.

These white blood cells produce antibodies which destroy viruses and bacteria.

Figure 2.28 The body's natural defence against disease

Antibodies are made inside the body by white blood cells. They fight infections caused by bacteria and viruses.

> ⑱ How are the pathogens destroyed after the antibodies have caused them to stick together?

> ⑲ Why does it take a few days before someone starts to get better after an infection?

> ⑳ Why is a small child unlikely to catch chickenpox a second time?

When someone has an infection, white blood cells respond by producing **antibodies** to fight it (Figure 2.29). This happens within a few days. The antibodies cause the pathogens to stick together. They are then ingested (taken in) by other white cells as shown in Figure 2.28. Once the antibodies have been produced, the sick person begins to get better. At the same time as antibodies are produced, 'memory cells' are also produced. These will produce antibodies rapidly if the same pathogen invades the body again so the person does not suffer from the same infection twice.

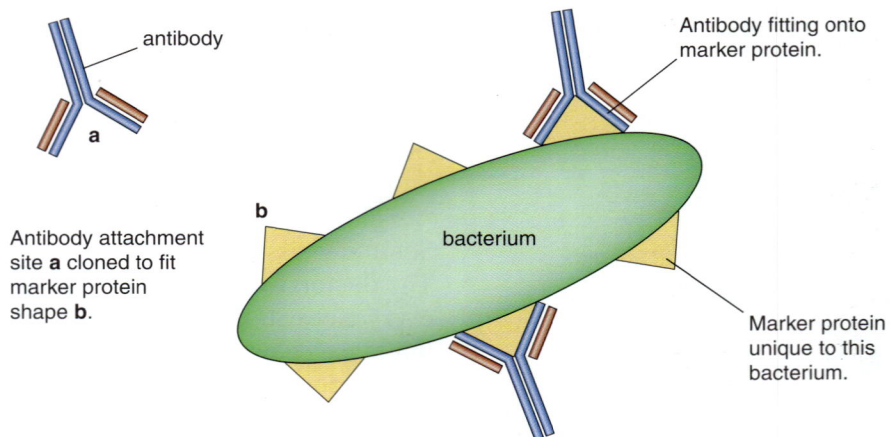

Figure 2.29 How an antibody fits onto the pathogen

2.7 Protecting our bodies against pathogens

Immunisation can protect our bodies against some pathogens. This involves injecting a solution which contains weak, inactive or dead forms of the pathogen. These weak forms of the pathogen may be dead bacteria, or part of a virus protein coat. The solution which is injected is called a **vaccine**. The white blood cells (shown in Figure 2.28) respond to the vaccine by producing antibodies which fit only that pathogen. At the same time, 'memory cells' will also be made. These memory cells make the person immune to further infection because they can produce the antibody needed to destroy the pathogen very quickly.

Vaccination can also be used to protect children against viruses such as those that cause measles, mumps and rubella (MMR). When viral vaccines are made, the genetic material is removed from the virus. Only the viral coat is used in the vaccine (see Figure 2.25). After vaccination, antibodies and 'memory cells' give the child lasting protection.

Antibiotics

Antibiotics are medicines which help to cure bacterial infections. They do this by killing the bacteria inside the body. Different antibiotics attack bacteria in different ways. The most common antibiotic is penicillin. This makes the bacterial cell wall porous. The bacteria then take in so much water that they burst. Other antibiotics prevent bacteria from reproducing. Some interfere with enzymes that the bacteria need to survive.

Our over-use of antibiotics has led to common bacteria becoming resistant to antibiotics. New strains of bacteria have resulted from a genetic change (**mutation**) in common bacteria. The mutation changes the chemical processes inside the bacteria in such a way that the new strain is resistant to most antibiotics (Figure 2.31).

Immunisation involves injecting a vaccine into someone to give them lasting protection against pathogenic bacteria or viruses.

A **vaccine** is a solution which contains weak, inactive or dead pathogens.

The HERD effect from vaccination

In the UK, we aim to vaccinate 95% of all children against common childhood infectious diseases. If this is achieved, the pathogen cannot be transmitted because there are few children left unprotected. Vaccinating most of the 'herd' protects the others.

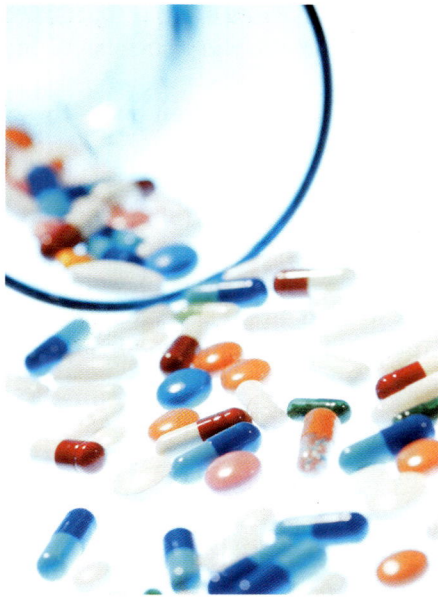

Figure 2.30 There are many different kinds of antibiotics.

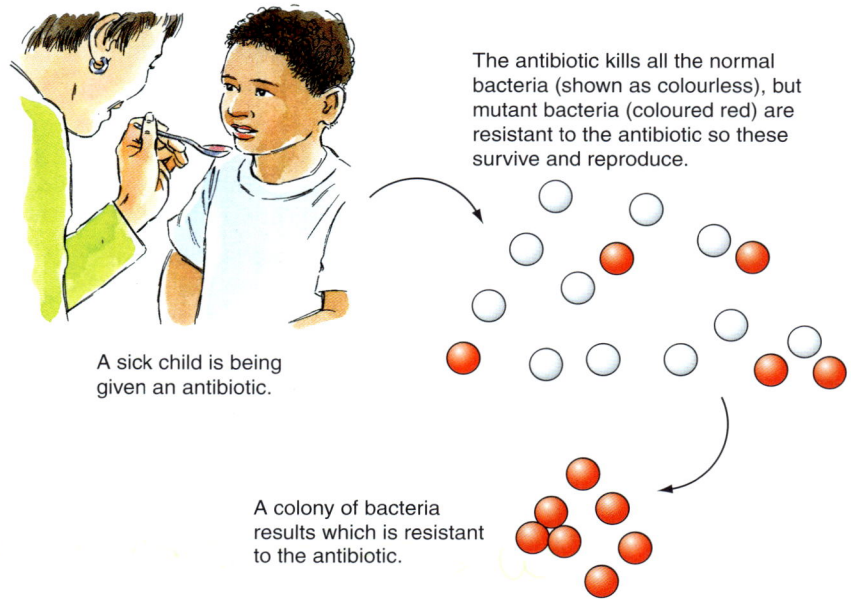

A sick child is being given an antibiotic.

The antibiotic kills all the normal bacteria (shown as colourless), but mutant bacteria (coloured red) are resistant to the antibiotic so these survive and reproduce.

A colony of bacteria results which is resistant to the antibiotic.

Figure 2.31 Development of bacteria resistant to antibiotics

You may have heard MRSA mentioned in the news. MRSA is a type of common bacterium which has changed genetically. It is resistant to most antibiotics. If a person with an infection is given an antibiotic, most bacteria will be killed, but any mutant bacteria may survive. Those that do survive will form a **colony** of bacteria resistant to the antibiotic. These resistant bacteria can be passed on to other people, especially in hospitals.

The best defence against passing on MRSA is careful hand-washing. This has led to a campaign in hospitals to remind staff and visitors to wash their hands before entering wards or touching patients. Hospital hygiene has improved but it needs to be watched carefully as bacteria are easily spread.

Viruses

Viruses cannot be destroyed using antibiotics. Viruses live and replicate inside cells. It is difficult to develop drugs which can enter cells and stop the viruses replicating without damaging the body cells. So we have to rely on our own immune systems to destroy viruses.

Many people are worried about 'bird flu' (avian influenza). If the genetic information (DNA) from a bird flu virus got inside a human flu virus, the new mutated virus could spread through the human population very quickly. This is what happened in the mild flu epidemics in 1957 and 1968. The serious flu epidemic in 1918 was a bird flu virus that adapted to humans.

At present, the bird flu virus is mostly in China and neighbouring countries. But the virus could spread around the world as ducks and geese migrate.

In developed countries, deaths from infectious diseases are rare because of **vaccination programmes**. Vaccination has proved very successful in the UK for the following reasons.
- There is a timetable for vaccination of babies and children up to 15 years of age.
- Most children are vaccinated.
- Parents are sent a reminder about regular vaccinations for their children.
- The vaccination programme is paid for by the government.
- There are enough community nurses to carry out the vaccinations.
- Most children are healthy and well fed so their immune systems respond to vaccines by producing antibodies and memory cells.

㉑ Copy the following paragraph and complete the blanks.

To help you complete the blanks, look at the six bullet points above. These explain why vaccination has been successful in preventing the spread of infectious diseases in the UK.

In the UK most children have the full course of _____ from the time they are babies until they are 15 years old. Parents are reminded by a _____ which comes in the post. In the UK, the _____ pays for vaccinations but, in less developed countries, the villagers or aid agencies have to pay. All vaccines have to be kept in refrigerators. Some poor villages in developing countries have poor roads and cannot be reached. The UK has many health centres with nurses but in developing countries there are not _____. Many children in developing countries are weak and poorly fed. They are not able to respond to vaccines by making a good supply of _____ and _____.

Summary

- A **balanced diet** should contain different food groups. It must provide essential amino acids, fatty acids, vitamins and minerals. It must also give the right amount of energy.
- A diet that lacks any of the above items leads to poor health.
- If you take in a larger amount of energy than you use, the excess is stored as fat.
- High levels of **cholesterol** in the blood increase the risk of heart disease.
- Cholesterol is transported round the body in the blood. Too much cholesterol is 'bad' because it can cause fatty deposits in blood vessels.
- **Metabolic rate** is the rate at which food is used by the body.
- Exercise raises the metabolic rate. It allows you to use up more food energy and keeps you slim.
- People who exercise regularly are usually fitter than people who take little exercise.
- In developed countries, people who are **obese** are likely to develop health problems such as arthritis, diabetes, heart disease and high blood pressure.
- In developing countries, most diseases are related to food shortage, dirty water and infectious diseases.
- Too much salt in your diet can increase the risk of heart disease.
- Processed 'ready-to-eat' meals often contain too much salt.
- **Pathogens** are **micro-organisms** which cause disease.
- The body's protection against disease includes the skin, mucus membranes and white blood cells.
- **Antibiotics** are medicines which kill pathogenic **bacteria** in the body.
- Bacteria reproduce rapidly and can undergo genetic changes (**mutations**). MRSA is a mutant type of bacterium which is resistant to many antibiotics.
- Hospitals can control the spread of bacterial infections by strict hygiene such as the washing of hands.
- In the nineteenth century, Semmelweiss observed that hand-washing reduced the spread of infection in a maternity hospital.
- **Immunisation** programmes are very good at protecting people against infectious diseases. **Vaccines** cause white blood cells to produce **antibodies** which destroy the infecting bacteria.

EXAMQUESTIONS

1 A diet with a large amount of red meat could result in:
A   too much calcium;
B   low body weight;
C   anaemia;
D   a high level of cholesterol in the blood.
(1 mark)

2 People who exercise are generally fitter than those who don't because:
A   they have a low metabolic rate;
B   they have small muscles;
C   their heart muscle pumps more blood per beat;
D   they have high blood pressure.   *(1 mark)*

3 Antibodies prevent infection by
A   causing bacteria to stick together;
B   sterilising the skin cells;
C   destroying the skin cells;
D   producing sticky mucus.   *(1 mark)*

4 Which one of the following statements about MRSA is **incorrect**?
A   MRSA is resistant to many antibiotics.
B   MRSA has developed resistance as a result of mutation.
C   MRSA can be stopped from spreading by good hygiene.
D   A person infected with MRSA cannot be cured.   *(1 mark)*

5 Which **two** of the following are causes of obesity?
A   Lack of exercise

B   Drinking too much water
C   Over-eating
D   Eating five servings of green vegetables each day   *(2 marks)*

6 A person eats some food contaminated with food poisoning bacteria.
a)   Which part of the body will these bacteria infect?   *(1 mark)*
b)   The person is ill for two days and then begins to feel better. Draw a diagram of the white blood cells that produce antibodies.   *(1 mark)*
c)   How do memory cells prevent a person catching an infection a second time?   *(2 marks)*

7 a)   How can doctors and nurses try to stop infections passing from one patient to another?   *(2 marks)*
b)   Write down two ways in which antibiotics can kill bacterial cells.   *(2 marks)*
c)   Explain how a bacterium can become resistant to an antibiotic.   *(3 marks)*

8 Write down three reasons why it is difficult to prevent the spread of an infectious diseases in a developing country.   *(3 marks)*

9 Give four reasons why trying to lose weight by eating only bananas is bad for you.   *(4 marks)*

10 Copy and complete the following table. The first line has been done for you.   *(8 marks)*

Choice A	Choice B	The best choice? A or B	Reason
White toast	Whole-wheat cereal	B	Whole wheat cereal has more fibre and releases sugar more slowly
Squash	Pure fruit juice		
Apple	Chocolate biscuit		
Jacket potato	Chips		
Chicken casserole	Chicken nuggets		
Jelly and ice cream	Fresh fruit salad		
Pot noodles	Curry and rice		
Glass of iced milk	Milk shake		
Burger and chips	Wholemeal ham roll with salad		

Chapter 3
What determines where organisms live?

At the end of this chapter you should:

✓ be able to identify the features that make some animals specially adapted to their habitat;

✓ understand how adaptations make some animals suited to cold environments and others to dry environments;

✓ recognise adaptations of some plants for dry environments;

✓ understand competition among plants and animals;

✓ understand the problems caused by more people using up more raw materials;

✓ be aware that more people use more land space;

✓ be aware that more people produce more waste and more pollution;

✓ have considered decisions that must be made to save the environment for the future.

Figure 3.1 Logging sites in rainforests damage habitats for wildlife.

3.1 Where do particular species live?

Look at the photo of the polar bear in Figure 3.2. How does it manage to survive in the Arctic?

Animals are said to be **adapted** to their environment when they are able to survive there and reproduce. To survive they must be able to get food from nearby. They must also be able to find a nest site or a place to feed and rear their young.

If you were going to the Arctic, what clothing would you take? Your packing list would probably include a warm fleece, a wind-proof hooded jacket, thermal gloves, and socks and boots.

Figure 3.2 A polar bear is well adapted to live in the freezing Arctic.

A polar bear can survive in cold conditions because it has:
- small ears, which lose very little heat;
- compact body shape to reduce heat loss;
- a thick layer of fur, which traps air and keeps heat in;
- a thick layer of fat – this keeps heat in and provides food during the winter when the bear is in a deep sleep;
- white fur, which camouflages (hides) the bear against the snow when it is hunting;
- white fur, which allows very little heat to be radiated away from the body.

Compare the features of a polar bear with those of a camel.

A camel can survive in the hot dry desert because it has:
- long legs and a long neck – the large surface area loses heat easily;
- thin hair on top of its body, which increases heat loss;
- no hair on the underside of its body, so it can lose heat here;
- little body fat, so heat can be lost easily from blood capillaries near the skin surface;
- a fatty hump which can be used to provide energy when food is scarce.

Figure 3.3 A camel is well adapted to live in the hot dry desert.

Camels have two rows of eyelashes. Their eyelashes and nostrils close for protection during sandstorms.

Plants which grow in very dry places have features to stop them losing too much water. The house leek (Figure 3.4) lives on rocks in areas where it is very dry in summer.

The house leek can survive without much water because it has:
- a short stem;
- fleshy green leaves which store water;
- a waxy, shiny coating to the leaves, which stops water evaporating;
- long roots which go deep between the rocks to find water.

These examples show that some animals and plants are specially adapted to survive where they live.

Figure 3.4 A house leek has fleshy leaves with a waxy coating.

Figure 3.5 An arctic fox lives in cold snowy places.

Figure 3.6 A desert fox lives in hot sandy places.

Look carefully at the photos of the two different foxes in Figures 3.5 and 3.6. Then copy and complete the sentences below.

1 The _____ does not lose heat easily because it has a thick fur coat which traps air. Air is a poor conductor of heat.

2 a) The arctic fox has _____ coloured fur in winter. This lets it get close to the animals it is hunting because they cannot _____ it.
  b) The desert fox, which lives in the sandy desert, has _____ coloured fur.

These animals are camouflaged in the places where they live.

3 The arctic fox has a thick layer of _____ under the skin to keep _____ in.

4 The arctic fox has short legs covered with thick fur but the desert fox has _____ legs covered with _____ fur.

5 Both foxes have very good hearing and use these for hunting. The ears of the desert fox are large so that it can _____ heat.

The loss of heat from different materials is studied further in Chapter 9.

3.2 ## What determines how many animals and plants live in an environment?

Plants need light and water to grow. The tallest plants get more light and those with the deepest roots get more water. Plants and animals that are trying to get the same food or live in the same space are **competing** with each other.

Primroses grow along hedgerows. They are small plants and cannot compete with trees for light. The leaves on the trees put primroses in the shade. So the primroses grow their leaves and flowers early in spring before the tree leaves open and put them in the shade.

Figure 3.7 These primroses flower early in spring before the trees have opened their leaves.

It is easier to see **competition** among animals than among plants. Figure 3.8 shows a crowded gannet colony.

The gannets compete for:
● a mate;
● a nesting site.

The gannet colony lives on a small rock. Flat places to make nests are hard to find. So the gannets fight for safe nest sites. Notice that the nests are placed 'pecking distance' apart. You can see why when you look at the sharp, pointed beaks of the gannets!

Figure 3.8 These gannet nests are evenly spaced apart. Each nest is just out of reach of the bird at the next nest.

Gannets are sea birds which feed on fish. They catch them by diving straight into the sea, beak first. If adult birds catch lots of fish, they will rear more young. But this will mean there is more competition for nest sites in the future.

Usually, there are gulls around a gannet colony. The gulls like to eat gannet eggs and young. Therefore gulls can reduce the number of gannets living in an area (the gannet **population**).

Three factors affect the number of gannets in a population:
1 competition for nest sites;
2 the number of fish available to feed the young;
3 predators which eat the eggs and young.

In the gannet colony most of the competition is among the gannets. Members of the same species are competing for food and nest sites. But there can also be competition between two different species that eat the same food. There are various ways to reduce this competition. For example, animals could feed at different times of the day. Hawks hunt for mice and frogs during the day. Owls hunt for mice and frogs at night.

Figure 3.9 The gannet has a very sharp beak.

Animals that catch and eat other animals are called **predators**. The animals that they catch and eat are called their **prey**. The gull is a predator. Its prey is the gannet.

6 What will happen to the number of mice in an area if a pair of owls are hunting to feed their young?

7 Name two other animals besides owls and hawks that help to control the mouse population.

8 Give two reasons why gardeners remove weeds from their flower beds.

9 Copy and complete the table by writing the names of the following animals in pairs in the correct columns, to show each predator and its prey.

| ladybird | mouse | cat | wasp |
| gannet egg | greenfly | seagull |
| caterpillar |

Predator	Prey

3.3 How do humans damage the environment?

There are 6000 million people in the world today, and the number is increasing.

How can we satisfy the needs of an increasing human population without damaging the environment?

Everyone wants a home, roads, shops, electrical goods and leisure facilities. And as the number of people increases, we need even more from the environment. In this section, we will look at the problems that this creates for a) land use, b) raw materials, c) waste, d) waste water treatment and e) air pollution.

a) Land use
More people mean that more land is needed for:
- homes, industrial estates, motorways and shopping centres;
- quarrying and mining raw materials for buildings and roads;
- large fields to grow crops;
- getting rid of rubbish (waste disposal).

An **ecosystem** consists of a habitat and the plants and animals living there.

Figure 3.10 This road carves through woodland, dividing the ecosystem.

A **biological control species** is a predator that eats a species that is regarded as a pest. For example, ladybirds eat greenfly. The ladybirds are the control species and the greenflies are the pests. Ladybirds reduce (control) the number of greenflies by eating them.

Figure 3.11 Ladybirds are a gardener's friend. They eat greenfly, which are pests.

A **brown field site** is an area in a town where old properties have been demolished. A **green field site** is farmland or 'green' land surrounding a town where there have been no buildings before.

A **finite resource** means that there is only a fixed amount available and it cannot be replaced. A **renewable** resource is one that will not run out.

How will increased land use damage the environment?

At present, between 10% and 20% of native plants and animals in the UK are in danger of dying out. The main reason for this is the increased use of land for homes, motorways and agriculture. Sites of Special Scientific Interest (SSSIs) are areas in which there are rare species that need special protection. The **ecosystem** around these plants or animals needs to be large enough to remain balanced and undisturbed. However, there can be problems if the rare species happens to be where a motorway is planned!

A **balanced ecosystem** changes slowly over time. The numbers of animals and plants will not change much. Animals need an area to find food and to reproduce. If the land area is divided up, the separate parts may be too small to maintain populations of larger animals such as deer and badgers.

When farmers remove hedges to make larger fields it can unbalance the ecosystem. Hedges and ditches provide food, shelter and breeding sites for animals. Some animals are predators, for example birds eat insects, ladybirds eat greenfly and wasps hunt for caterpillars to feed their larvae. We call these predators **biological control species** because they keep down (control) the numbers of species that are pests. If the predators have nowhere to live, the pests can get out of control.

One way of leaving more land for farming and recreation is to landscape and develop areas inside towns such as those occupied by old industrial buildings or derelict houses. These areas are called **brown field sites**. One big advantage of using brown field sites is that new roads don't have to be built. Some brown field sites have been used to make shopping centres.

b) Raw materials
Important raw materials like building stone, metal ores and fossil fuels are **finite resources**. If we use more of these, they will run out sooner. Alternatives then have to be developed such as **renewable** energy sources and recycled building materials. Trees are grown and the wood is used as a building material. But, as trees are cut down, more trees must be planted to replace them. There is more information about our use of raw materials in Sections 5.1 and 5.6.

c) Getting rid of waste
We get rid of most household and industrial waste by putting it in areas of land such as disused quarries. These are called **landfill sites**.

Waste must be managed properly. Otherwise there can be problems with pollution. The environment can also be damaged. Waste tips can be unsightly. Bacteria from decomposing material can be carried by birds such as gulls, which feed on the waste. Toxic chemicals can be washed into rivers by heavy rainfall.

Figure 3.12 Landfill sites attract scavengers such as gulls.

Waste disposal is a difficult problem. New solutions need to be found.

Some facts:
- The UK produces over 100 million tonnes of waste per year.
- As the number of people increases, the amount of waste also increases.
- Most rubbish is dumped in landfill sites. These are becoming full and few other sites are available.
- A lot of the material in landfill sites could be recycled. This would save raw materials.
- The Government aims to reduce landfill waste to 55% of the present amount by 2010.

So what can we do with our rubbish? We could recycle more of it. We could burn more of it in incinerators. We could turn more of it into compost. But, it would be expensive to build the processing plants to do this. However, some of the money could be recovered by selling the recycled materials.

In many areas of the UK, some of our rubbish is collected for **recycling**. Local councils encourage people to separate paper, glass, metal and plastics from their rubbish. This is collected and sent to companies which can recycle these materials (see Section 3.6).

Incineration is a process of controlled burning. After removing the materials that can be recycled, the remaining waste is incinerated. This leaves only a small amount of ash. The ash can be used for road building instead of natural stone. All new incinerators are Energy from Waste (EfW) plants. This means that the heat produced from burning the rubbish is used to make electricity. People don't want incinerators to be built near their homes. They believe that burners will produce smoke and smells, as they did in the past. But modern incinerators are carefully controlled so that gases passing out of the chimney are at a safe level.

Garden waste such as weeds, cut grass and leaves can be collected for **composting**. These materials are shredded and left to decompose (break down) into compost. The compost is then treated to kill off any harmful bacteria and sold.

d) Waste water treatment
As the number of people increases, so does the amount of sewage they produce. Waste water (sewage) is treated in special processing plants to remove toxins (poisons). Then it is piped into rivers or the sea. Waste water treatment has had to improve to treat more and more sewage. If the sewage is not treated properly it will pollute river water. This can sometimes cause the river to turn green and become clogged up with plant growth (**eutrophication**).

---

⑩ Name four materials which are collected from homes in many areas for recycling.

⑪ a) Name four ways of getting rid of waste.
   b) Which of the methods will reduce the amount of waste?

⑫ Do you think people are right to be worried about an incinerator being built near their homes? Give reasons.

Eutrophication is caused when nutrients are added to water in a river. The nutrients may be nitrates and ammonium compounds from fertilisers, or phosphates from detergents. As a result, algae in the water reproduce rapidly and the water becomes green. Other water plants also grow rapidly and can block the river.

Figure 3.13 Filtering water through reed beds – an opportunity for a new ecosystem

Activity – Campaign poster

A local council is planning to build a new road through the middle of an area of old woodland. This woodland contains rare plants, rare butterflies and a few deer. It is also a popular area for walkers at the weekend. Make a poster to show why the road should be built around the wood and not through it. Use illustrations to add interest.

Most water pollution happens after heavy rain. The heavy rainfall floods the pipes from the sewage works and this can carry raw sewage into rivers.

Many chemical companies and power stations now use biological methods to remove nitrates and toxins from their waste water. They filter the waste water through beds of reeds. The reeds use nutrients in the waste water to help them grow and also remove toxins from the water. In addition, the reed beds provide a wildlife habitat for many birds and insects.

e) Air pollution

Most air pollution comes from the exhaust gases of vehicles on our roads. In town centres we breathe in traffic and diesel fumes affecting our health. Factories, power stations and bonfires also pollute the air in our towns and cities. If the fuels used contain sulfur, the waste gases will contain sulfur dioxide. This causes acid rain which can even damage stone in buildings.

At the beginning of this section, we asked the question 'How can we satisfy the needs of an increasing human population without damaging the environment?' After examining different issues, we can make some positive suggestions.
- Avoid damaging balanced ecosystems.
- Use brown field sites in town to build new homes. This saves the countryside and farmland and means there is no need for new roads.
- Increase the recycling of paper, glass, metals and plastics which will save raw materials.
- Use biological methods to treat waste water.

13 As the number of people increases, we use more transport. State two ways in which this can affect our ecosystem.

14 The insects shown in Figure 3.14 eat pests which damage plants.

a) What is the name given to these useful insects?

b) Why have the numbers of these useful insects reduced?

Figure 3.14 Hoverflies, ladybirds, wasps and ground beetles

3.4 Deforestation – short-term gain, long-term loss

Huge areas of tropical rainforest are a very valuable resource for the world. They provide many useful plants which are used to produce drugs to treat cancer. Their trees also provide a lot of the world's supply of oxygen as a result of photosynthesis.

Figure 3.15 A small tribal village clearing in the rainforest

The ecosystem of a rainforest depends on nutrient recycling. Dead leaves on the forest floor decompose and add nutrients and minerals to the soil. This makes the soil fertile. Tribal villagers clear small areas within the rainforest to use to grow crops. But the thin soil remains fertile for only a few years. So the villagers move to a new site. Gradually forest will grow back into the small cleared areas that have been left.

However, the trees from rainforests are also useful. When trees are cut down, the timber can be sold for money or used as a building material. Large areas of cleared land can be used for cattle ranching or to grow crops for sale. These activities provide much needed income for the country. Cleared land can also be used to build roads connecting major cities. This helps in selling the goods. Many people therefore want to cut down the trees. Logging companies have destroyed large areas of the rainforest in this way (see Figure 3.1).

When large areas of trees are cut down (large-scale **deforestation**), problems result:
- There are huge cleared areas in which the forest cannot grow back.
- The trees normally break the heavy rainfall. When the trees are removed the rainfall is much more damaging to the land.
- The heavy machinery used to remove the trees causes damage to the soil and roots.
- Soil erosion occurs when heavy rainfall flows across the cleared land. It washes away the soil on the surface.
- When soil is washed into streams and rivers, they become blocked.

Large-scale deforestation also causes other problems:
- Animals that live in parts of the rainforest become isolated because they cannot cross the large cleared areas.
- Destroying the forest reduces the supply of food for animals and local human populations.

Huge areas of deforested land in the Amazonian rainforest have already been abandoned. The thin soil cannot support crops or cattle for more than a few years and these areas are slowly becoming deserts.

15 What are the quick gains from deforestation?

16 Why does the soil only support crops for a few years?

17 How does soil erosion take place?

18 What is the long-term damage caused by deforestation?

19 It has been suggested that narrow blocks of rainforest should be cut down with mature rainforest left on each side. What advantages would there be in this type of management?

The Brazilian Government wants to save the rainforest because it is such a valuable resource to the world. But the Government also wants to make life better for the increasing population. It has to decide how to manage the rainforest resource for the best.

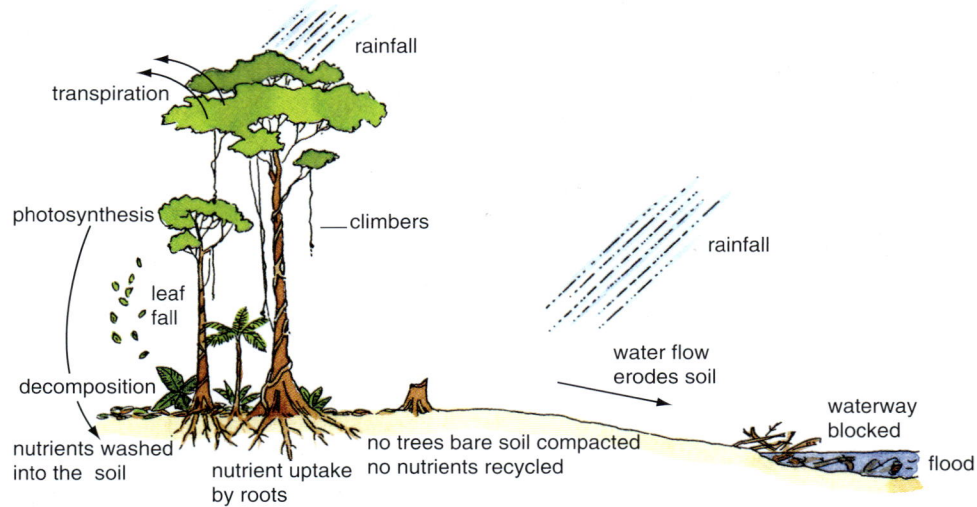

Figure 3.16 The rainforest cycle and the results of cutting down the trees

How does cutting down the rainforest affect the atmosphere?

The rainforest is a **balanced ecosystem**. In an undisturbed rainforest there is a balance between the carbon dioxide being removed from the atmosphere and that being released.

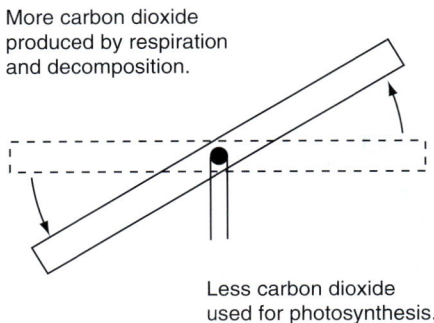

Carbon dioxide removed from atmosphere by trees during photosynthesis **=** Carbon dioxide released into atmosphere by bacteria and fungi which decompose dead trees and plants **+** Carbon dioxide released into atmosphere by respiration of animals and plants

Logging causes more carbon dioxide to be released into the atmosphere. The carbon dioxide come from:
● burning the waste branches in huge fires;
● increased decomposition as bacteria and fungi break down all the dead plant material.

Trees 'lock-up' carbon from carbon dioxide in compounds such as cellulose (wood), as they grow over many years. The balance in the rainforest is suddenly upset because all the carbon that was 'locked-up' in cellulose is released in one massive fire. Slow re-growth of trees and plants means that not much photosynthesis can take place. So the carbon dioxide stays in the atmosphere.

More carbon dioxide produced by respiration and decomposition.

Less carbon dioxide used for photosynthesis.

Figure 3.17 Deforestation increases the amount of carbon dioxide in the atmosphere.

3.5 Global warming

A **greenhouse gas** is a gas such as carbon dioxide or methane. These gases trap heat in the Earth's atmosphere.

Scientists have found an unusual link between deforestation, cattle ranching and global warming. Deforestation provides land for cattle ranching. Cattle eat grass. They have anaerobic bacteria in their stomachs, which break down the cell walls in grass. These bacteria produce methane gas, which the cows release into the air. Unfortunately, methane is a powerful **greenhouse gas** which adds to global warming.

The anaerobic bacteria in cattle are not the only methane producers. Methane is also produced in rice-paddy fields by anaerobic bacteria which decompose vegetation in water-logged conditions. As the population of Asia increases, more rice is needed. So more land is flooded, increasing the amount of methane released. Methane is also produced by the anaerobic decay of materials in sewage and household refuse. In some countries this methane is collected, stored and used as fuel.

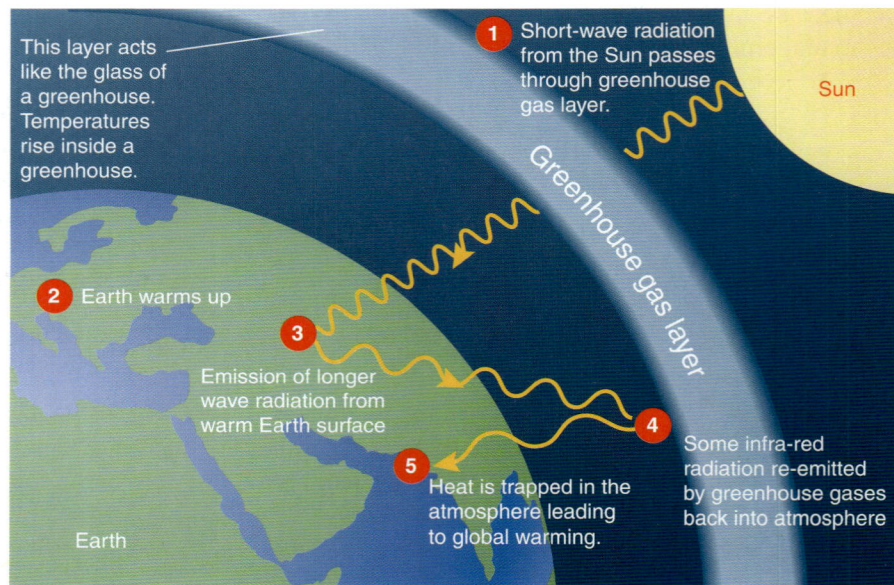

This layer acts like the glass of a greenhouse. Temperatures rise inside a greenhouse.

1 Short-wave radiation from the Sun passes through greenhouse gas layer.

Sun

Greenhouse gas layer

2 Earth warms up

3 Emission of longer wave radiation from warm Earth surface

4 Some infra-red radiation re-emitted by greenhouse gases back into atmosphere

5 Heat is trapped in the atmosphere leading to global warming.

Earth

Figure 3.18 The greenhouse effect

How do scientists investigate global warming?

At present, scientists are trying to find out if global warming is caused by an increase in greenhouse gases. They have two pieces of crucial evidence:

- concentrations of greenhouse gases in the atmosphere are increasing;
- temperatures around the world are getting hotter.

The Earth has gone through periods of global warming before. So the warming that we see now may not be linked to greenhouse gases, although many scientists believe it is.

To test their ideas, scientists produce **scientific models** using computer programs which can take account of many factors. If the model works for temperature data from the past, the same model can be used to predict future temperature changes. Models like this are used in weather forecasts.

Here are some other observations related to global warming.
- The Arctic winter has become shorter and the ice caps are melting.
- Sea levels have risen between 10 cm and 20 cm in the last hundred years. This is because more snow has melted in the northern hemisphere.
- Graphs of temperature and carbon dioxide levels for the last 100 years show similar increases.
- In recent years there has been unusually heavy rainfall and flooding in the UK and northern Europe.
- During the same period, droughts and crop failures in the Sudan and other African regions have increased.

These observations fit neatly with the predicted effects of global warming from scientists using their computer models.

Should we be worried about global warming?

The present rise in worldwide temperatures is faster than ever before. It seems that scientists have under-estimated the rate at which world temperatures would rise. If temperatures continue to rise and the ice caps continue melting, sea levels will rise further. Areas of Holland, Bangladesh and eastern England which are only just above sea level will be flooded. In time, rising temperatures could change the direction of rain-bearing winds. This would cause droughts in some countries and flooding in others. Some people think that this has already started.

Governments throughout the world should try to reduce the production of greenhouse gases. In the next section, we will look at ways in which we can all help.

3.6 What can each of us do to reduce damage to the environment?

Recycling waste

We can all help to reduce damage to the environment by recycling our waste. People who recycle waste materials are helping to save natural resources. Everyone must help to achieve the local and national targets to reduce waste. Local authorities are making recycling easier.

There are bottle banks and facilities for recycling aluminium cans. Many towns also have kerbside collections for waste paper, bottles and cans. Countries such as Germany have been doing this for longer than the UK. They have reduced waste significantly.

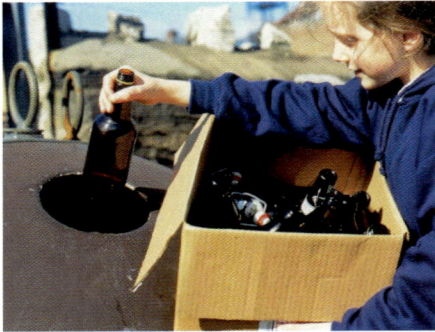

Figure 3.19 More young people are recycling and trying to reduce waste.

Activity – Waste not, want not!

❶ Find out about your local authority plan for reducing waste. You should be able to find this information on their website under 'Waste – Local Plan'. What are the key points in your local plan?

❷ Check out the facts on waste by visiting the website of the Environmental Services Association for waste facts: www.esauk.org/waste/facts

❸ If your town has kerbside collections, find out which materials are collected separately.

❹ Make a list of six common products that can be produced from recycled materials. If you need help, go to www.recycledproducts.org.uk which has lots of interesting links.

Making our homes more energy efficient

We can also prevent damage to the environment by making our homes more energy efficient. We can install loft insulation, cavity-wall insulation and double glazing (see Section 9.4) to stop heat escaping from our homes. By doing this:

- less fuel is needed and the reserves of fossil fuels last longer;
- less carbon dioxide is released into the atmosphere.

⓵⓽ List four other ways in which you could reduce the energy used at home or at school. Visit www.est.org.uk to get some ideas.

Walking or cycling

Walking or cycling instead of driving is another way in which we can reduce damage to the environment. Walking and cycling do not use petrol and diesel. This:

- reduces air pollution;
- saves petrol and oil, which are finite resources.

Walking and cycling are also ways of maintaining a healthy weight and keeping fit (Chapter 2). Using cycle routes may even save you time!

Figure 3.20 Where is your nearest cycle route?

Activity – What can you do for a better future?

1 Work in a small group. Develop a plan that reduces waste or increases recycling. It can be a plan for the community in which you live or a plan for your school. If you need help go to www.sustainable-development.gov.uk or www.rspb.org.uk/green/index.asp for some ideas.

2 Make a poster, webpage or PowerPoint presentation to show your plan to the rest of your class. Remember to explain the importance of each suggestion in your plan. For example, recycling aluminium cans saves the raw material and saves the energy used in refining aluminium ore.

3 If possible, get permission to put your plan into action. Remember you are more likely to keep up simple tasks, such as collecting cans or organising a petition.

It is important to improve the quality of life now and in the future. This means saving raw materials, saving valuable resources and keeping a good clean environment for us, our children and grandchildren. This need to improve life not just now but also for future generations is called **sustainable development**.

3.7 What is our own government doing for sustainable development?

The UK government has already produced a plan for sustainable development. This plan supports living 'within the means of the environment' so that natural resources are available for future generations.

Part of the plan is to limit fishing so that fish grow to be adults which produce offspring. This will allow fish populations to increase again. If over-fishing continues, some fish could become extinct.

A second part of the plan encourages the redevelopment of run-down sites within towns (brown field sites) rather than using up farmland or parkland.

For further information on the government's plan go to www.sustainable-development.gov.uk.

In 1992, more than 100 countries met for the first International Earth Summit. A major plan was suggested for sustainable development. Its aim was to reduce poverty and give people access to the resources to support themselves. The plan asked richer countries to lead the way by reducing pollution, reducing emissions of greenhouse gases and reducing the use of natural resources. 'Make poverty history' was the slogan in 2005 but unfortunately little progress has been made.

Activity – Easter Island – a case of unsustainable development

Figure 3.21 Easter Island was once a wooded island.

Easter Island in the South Pacific Ocean is known for its amazing statues. These statues were carved about 1200 years ago. Some of them are up to 10 m high. The island was heavily wooded when the first inhabitants arrived 1300 years ago. The palm trees on the island were ideal for building houses and hollowing out for canoes. Canoes were used for catching seafood, especially porpoises. The statues were carved from volcanic rocks. But these rocks were a long way from the place where the statues were finally placed. How did the people move the statues? They rolled the statues on huge logs and pulled them with ropes made from climbing plants.

The population on Easter Island increased rapidly and the forest was cleared as the people cut down trees for more buildings. Without the protection of trees, the thin soil was blown away. Crops failed. There were no trees left to make boats. After the people had eaten all the porpoises, seabirds and shellfish, they began to starve. Fighting and cannibalism broke out. Eventually, no one was left. At one time the island was rich in forests. Today there are no trees on Easter Island.

Sadly, there is evidence of similar damage to environments in the Brazilian rainforests, in parts of Africa and in Asia. This is why a world plan for sustainability is needed before it is too late.

❶ Copy and complete Table 3.1. It compares the disaster on Easter Island with tribal use and commercial logging in the Amazon rainforest.

	Easter Island	Tribal use of rainforest	Commercial logging in rainforest
Timber used for:	Building homes Building boats/canoes Rollers for moving statues Fuel		
Main food source:		Bush animals Fruits, berries and roots Grubs	Food is delivered to logging camps and results in food waste and packaging waste
Soil erosion a problem?	Yes – because trees were removed		
Effect on animal populations:	Reduced by over-fishing and habitat destruction	Little damage – small areas used in which plants regenerate quickly and animals spread back from nearby areas	
Problems faced by humans:	Starvation as a result of over-fishing and habitat destruction		Conflict between displaced tribes and loggers
Can the environment be restored?	Not naturally. There has been damage to the soil and no seeds from native plants to re-grow.		

Table 3.1 Comparing the disaster on Easter Island with tribal use and commercial logging in the Amazon rainforest

3.8 Using plants and animals to detect levels of pollution

When scientists measure levels of pollution they usually analyse gases in the air or chemicals in river water and sea water. Their results tell them what is in the sample. Chemical analysis like this gives no information about pollution in the past. But the pollution may have killed organisms, leaving clues for a biological detective.

Using lichens as air pollution indicators

Lichens (pronounced 'li-kens') grow as crusty patches on rocks and tree trunks. Lichens consist of a fungus thread and a single-celled plant which live together. Different species of lichens have different sensitivities to sulfur dioxide. Some lichens are so sensitive to sulfur dioxide that a trace of the gas in the air will kill them. This means that by looking at the lichens growing in an area we can work out how much pollution from sulfur dioxide is in the air. Because of this, lichens can be used as **biological indicators** (**bioindicators**) of pollution.

Information from lichen surveys can be used to record the pollution caused by increased numbers of vehicles along a road. This can help in decisions about where new roads or industrial sites will cause the least pollution.

Lichen 1	Lichen 2	Lichen 3	Lichen 4
This common orange lichen is found on trees and walls. It is able to survive in areas with some sulfur dioxide pollution.	This crusty grey lichen is common on tree trunks. It survives in areas of low sulfur dioxide pollution. It dies quickly if pollution levels rise.	A common woodland lichen. It can only survive in clean air.	This is often called beard lichen. It grows very slowly. It only survives in pure air with extremely low levels of sulfur dioxide pollution.

Table 3.2 Common lichens which act as bioindicators of air pollution

Activity – A pollution survey

A science group decided to survey the air quality around their town using lichens as bioindicators.

They were given the four pictures of lichens in Table 3.2. These lichens can be used as pollution bioindicators.

Sulfur dioxide is produced when fossil fuels are burned. So the concentration of this gas is high near main roads, and in residential and industrial areas. With these points in mind, the students made the following predictions.

Predictions:
i)   Only the tolerant orange lichen will be found in areas near main roads, houses and industrial units.
ii) Crusty grey lichen will occur in areas where there is not too much traffic.
iii) Woodland lichen will be found on trees well away from roads and industrial areas.
iv) Beard lichen will only occur where the air is pure.

Before starting their survey, the students made a copy of the town map. On it they marked:
- industrial areas in blue;
- main roads in red;
- parks and woodland in green.

Winds in the town blow from the south-west most of the time. So any sulfur dioxide will be blown towards the north-east of the town.

The students drew eight lines along the compass bearings N, NE, E, SE, etc. from the town centre on the map. They also drew concentric circles around the town centre, each 0.5 km apart up to a radius of 5 km. Survey points were chosen where the compass lines and circles crossed. At these survey points, students looked for lichens similar to the four indicator species shown in Table 3.2. They recorded the lichens seen on the nearest walls, trees and hedges at each survey point. When the students got back to school, their results were plotted onto the map.

Figure 3.22 A map showing the results of the lichen survey

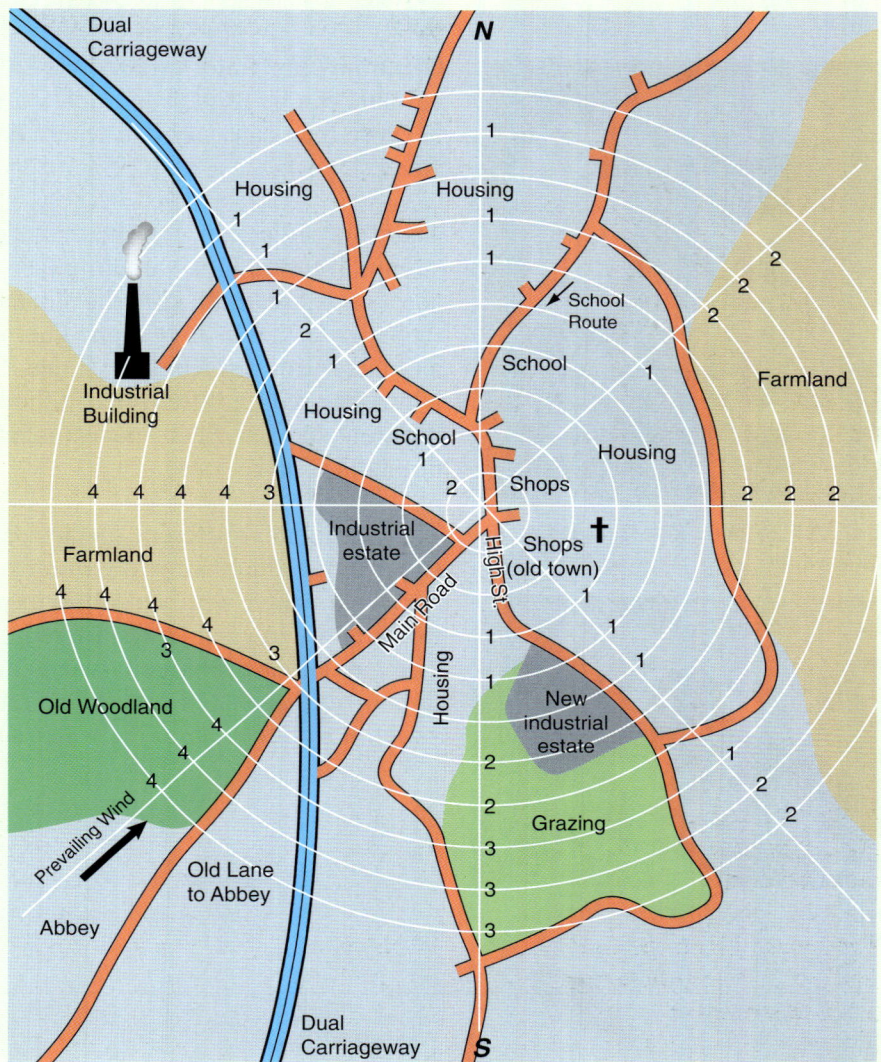

❶ Use information from the survey and the map to copy and complete the following statements.
a) Different lichens have different sensitivities to _____ dioxide, which is produced when _____ fuels are _____. The wind blows this gas to the _____ side of the town. The air on the _____ side of the town will be cleaner.
b) Beard lichen is very sensitive. It grows very _____. The Abbey ruin and nearby wood are very old and undisturbed. The beard lichen growing here shows that the air is _____.
c) In the 2 km radius around the town centre, the main lichen is _____. This shows that there is a _____ level of _____ _____ pollution.
d) The only specimen of lichen 2 near the town centre was found in the old churchyard. Lichen 2 was mainly found _____.
e) Lichen 3 indicates _____ air. The closest point that it was growing to the town centre was _____ km on the _____ side of town.
f) North-east of the industrial building on the west side of town the students found lichen _____. They decided there were two possible reasons for finding this lichen there. Either the smoke from the industrial building contained _____, or the area was quite close to _____.

❷ Were the students' predictions supported by their results? Say 'yes' or 'no' and explain your answer.

Using animals as bioindicators of water cleanliness

Clean flowing water usually has lots of oxygen dissolved in it. Some animals can be used as bioindicators for water cleanliness. This is because they can only live where there are high levels of oxygen in the water.

The usual procedure for checking the pollution-sensitive species in a stream involves kick-sampling (see Figure 3.23). The investigator wades across the stream kicking the stones to disturb the small animals and then crosses back. A D-shaped net is held downstream to catch the animals dislodged from the stones. The catch is then emptied into a white tray with a little water in the bottom. The animals are sorted into different species. Each species is recorded as rare, frequent or in large numbers (Figure 3.24).

There are many different species of wildlife in a clean well-oxygenated stream. A polluted stream will only have a few species. These can survive in water with very little oxygen in it. Two species which can tolerate low oxygen levels are red midge larvae and rat-tailed maggots. The red colour of midge larvae is haemoglobin which picks up oxygen in poorly oxygenated water. The rat-tailed maggot has a breathing tube which it stretches up to the water surface to take in air.

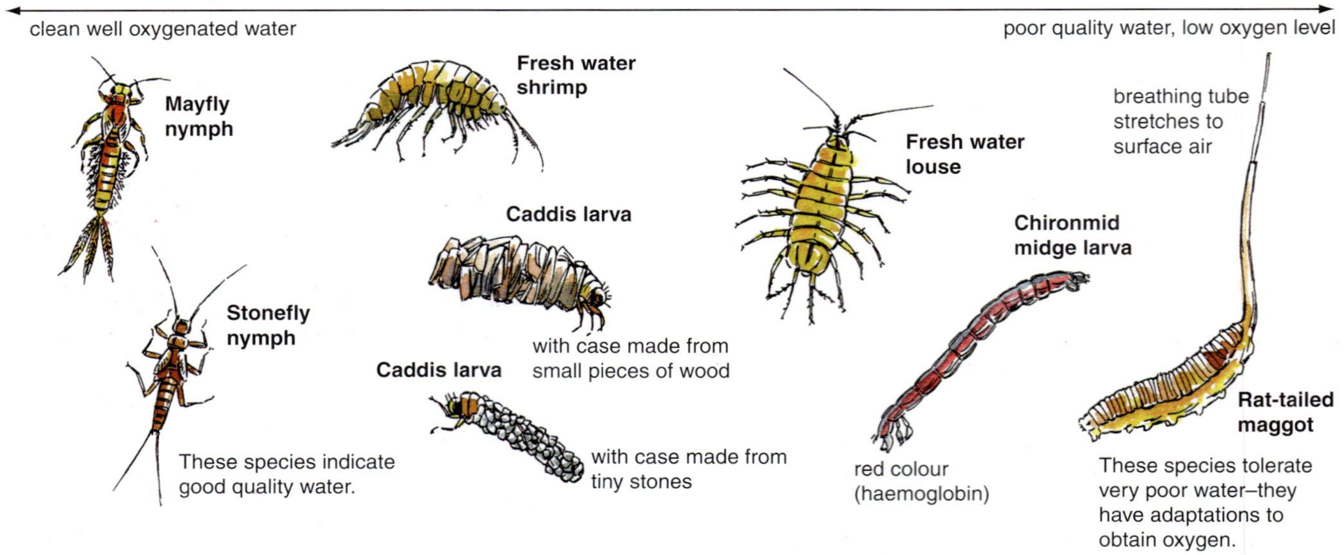

Figure 3.23 Students kick-sampling in shallow water

Figure 3.24 Sorting the kick-sampling catch

clean well oxygenated water → poor quality water, low oxygen level

Mayfly nymph

Fresh water shrimp

Fresh water louse

breathing tube stretches to surface air

Caddis larva

Chironmid midge larva

Stonefly nymph

Caddis larva

with case made from small pieces of wood

with case made from tiny stones

red colour (haemoglobin)

Rat-tailed maggot

These species indicate good quality water.

These species tolerate very poor water–they have adaptations to obtain oxygen.

Figure 3.25 Animals that indicate the quality of the water they live in

Summary

✓ Animals must obtain all they need to eat, grow and reproduce from the environment in which they live.

✓ Animals have features called **adaptations** which allow them to survive in the environment in which they live. These features enable them to survive at the temperature of the area, find food within the area and locate a breeding site.

✓ **Population size** is controlled by **predation** and **competition** for food, water and breeding sites.

✓ Plants growing in dry areas have adaptations to reduce water loss, such as thick waxy cuticles, water storage tissue and long roots.

✓ Plant populations are controlled by the numbers of herbivores (grazing) and by competition for light, water and minerals.

✓ A **balanced ecosystem** changes slowly and the numbers of producers, herbivores and carnivores remain fairly constant.

✓ **Biological control organisms** such as ladybirds, insectivorous birds and ground beetles control pest species, such as aphids. Kestrels and owls control mice.

✓ The increasing human population has an impact on the environment through the use of more land for buildings, roads, intensive farming and waste disposal.

✓ The increasing human population uses more water and raw materials.

✓ The increasing human population produces more air pollution, more water pollution, and more waste.

✓ **Deforestation** for timber production, arable farming and cattle ranching can have damaging effects on the whole environment including the land, the wildlife and the human population.

✓ **Global warming** is the result of increased carbon dioxide and methane levels in the atmosphere.

✓ Conclusions linking global warming and greenhouse gas levels require careful measurements over many years, not anecdotal evidence.

✓ In environmental situations where many variables are possible, scientists construct computer models to manipulate the data and make predictions for the future. This applies to predictions about global warming.

✓ The consequences of global warming are climate changes which result in the flooding of low-lying areas and change in the direction of rain-bearing winds.

✓ Action taken by the public to reduce waste, recycle more, and use energy efficiently will benefit society and the environment.

✓ **Sustainable development** is necessary to make sure that fuel, food and a pleasant environment are available for future generations.

✓ Environmental monitoring can be carried out through chemical sampling, but **bioindicators** of water or air pollution give a longer term view of changes.

✓ Sampling of bioindicators requires a well-planned investigation and data collected over many years for comparison.

EXAMQUESTIONS

❶ Copy and complete the sentences below using the words from this box.

nitrates	sand	toxins	flooding
nutrients	soil	tins	eutrophication
	landslides		

a) If heavy rainfall washes over a landfill site _____ could be washed into a river and kill fish.

b) If heavy rains flood a sewage works _____ could be washed into a river causing _____.

c) Heavy rains falling on a deforested hillside could wash away the _____. If this goes into a river it could cause _____. (*5 marks*)

❷ Write the answer 'true' or 'false' in your notebook for each of the statements a) to e) below.
   a) An EfW incinerator produces thick black smoke.
   b) Ash from an incinerator can be used to make roads.
   c) An EfW incinerator produces heat which can be used.
   d) Incinerators are not controlled and produce poisonous gases.

EXAMQUESTIONS

e) Incinerators reduce the waste to a small volume of ash. *(5 marks)*

❸ Give three reasons why landfill is an unsatisfactory method of getting rid of waste. *(3 marks)*

❹ Copy this simple drawing and complete the labels.

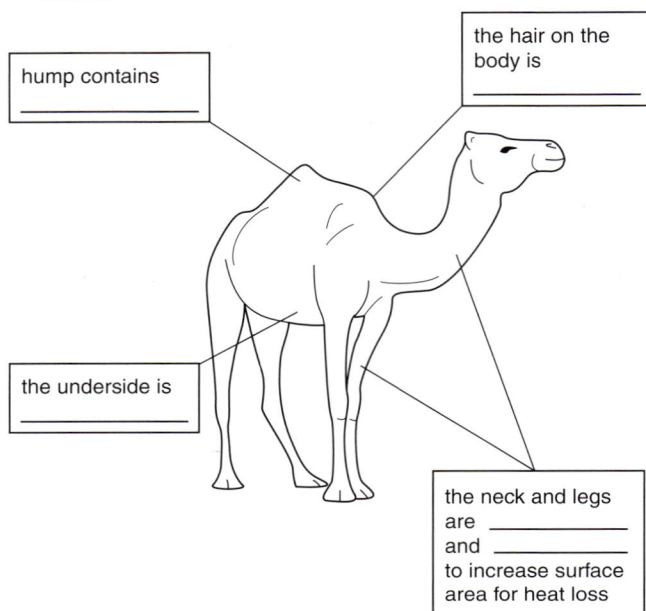

the hair on the body is _____

hump contains _____

the underside is _____

the neck and legs are _____ and _____ to increase surface area for heat loss

Figure 3.26

❺ West Brook is a shallow stream. Four groups of students decided to investigate the small organisms in the water there. The technique of kick-sampling was used.

Groups A and B sampled upstream from a row of cottages. Here the water was clear and the stones on the bottom could easily be seen.

Groups C and D sampled downstream from the cottages. Here the water was less clear and the stones were covered with a black slimy deposit.

Figure 3.27 shows a map of the survey sites.

The results of their samples are shown in Table 3.3.

Choose the correct answer from A, B, C and D in each of the following four questions.

a) The results show that:
A twice as many specimens were caught upstream compared with downstream
B more species were caught upstream than downstream
C the least sensitive species were found upstream
D rat-tailed maggots were very common *(1 mark)*

b) The group said that the water upstream was unpolluted because:
A few red midge larvae were found there
B a small fish was caught there
C many different species were found there
D the number of any species was small *(1 mark)*

c) What could have happened by the cottages to change the wildlife in the stream?
A sewage may have seeped into West Brook
B fertiliser may have been washed off the fields
C trees on the banks may have caused shading
D more fish may have eaten the invertebrates *(1 mark)*

d) Which of the following would **not** improve the kick-sampling results?
A inspect the net before use to make sure it is completely empty
B sample for the same period of time in each location
C always hold the net on the upstream side
D kick carefully from the bank on one side to the other side and back *(1 mark)*

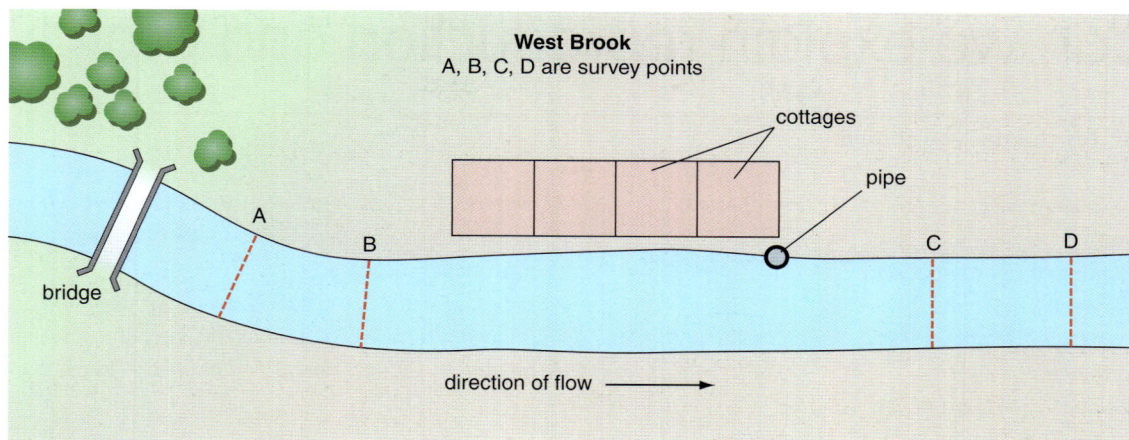

Figure 3.27 Map of survey sites and cottages

Organisms	Group A	Group B	Group C	Group D
Stonefly nymphs	+++++	+++++		
Mayfly nymphs	+++	+++++		
Dragonfly nymph		+		
Small fish	+			
Freshwater shrimps	+++++	+++++		+
Freshwater lice	+++	+++	+++	+++
Caddis with vegetation case	+	+		
Caddis with stone case		+		
Daphnia	+++++	+++++		
Water snails	+++++	+++++		
Water boatman beetle	+++	+		
Small worms	+++++	+++++		
Midge larvae (red)	+	+	+++++	+++++
Rat-tailed maggot			+++++	+++++
Total number of species				

Key: + indicates one specimen, +++ several, +++++ large number.
Table 3.3 The results of a survey of West Brook stream

EXAMQUESTIONS

Chapter 4
How can we explain reproduction and evolution?

At the end of this chapter you should:

✓ be able to explain how genes are passed from parents to their children;
✓ know that genes carry information;
✓ know where genes are found;
✓ be able to describe the differences between sexual and asexual reproduction;
✓ know how clones can be produced;
✓ be able to judge the economic, social and ethical issues concerning cloning and genetic modification;

✓ be able to compare different theories about evolution;
✓ be able to explain why Darwin's theory of evolution is the most widely accepted;
✓ know how fossils provide evidence for evolution;
✓ understand how evolution occurs via natural selection;
✓ be able to suggest why different species became extinct.

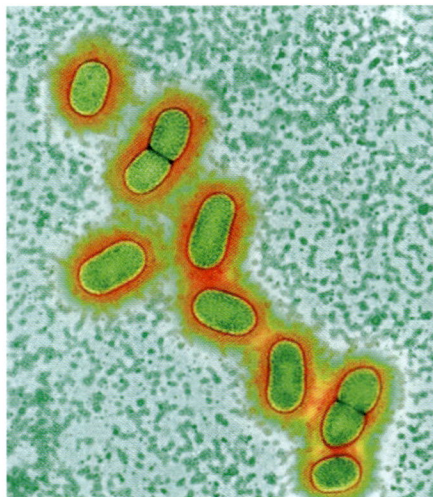

Figure 4.1 In all these photos, characteristics have been passed on from one generation to the next.

4.1 How did an Austrian monk help our understanding of inheritance?

Inheritance means the passing on of appearance and characteristics from parents to their offspring.

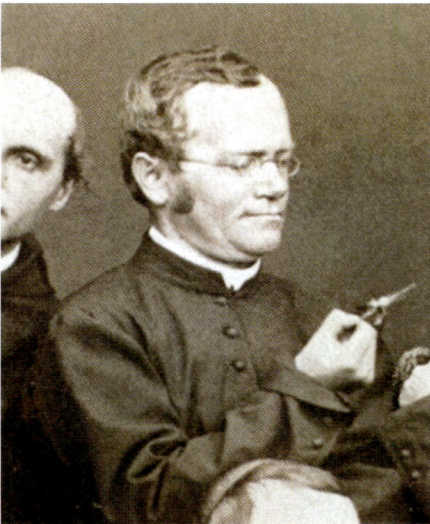

Figure 4.2 Gregor Mendel at work

Your looks are partly passed on to you from your parents. This process is called **inheritance**. In this section, you will learn about the key developments that led to our understanding of inheritance.

For many years, people thought that a mother's and father's looks were combined when they had children. For example, a black haired mother and blonde haired father would have children with brown hair. This was known as the 'blending theory' of inheritance.

An Austrian monk called Gregor Mendel, born in 1822, disagreed with the blending theory. Mendel worked in the gardens of a monastery. He noticed that pea plants always had either purple or white flowers. Seeds from both types of plants grew into plants with purple or white flowers. But they never grew plants with pink flowers, a mixture of the two colours. This is why Mendel did not agree with the blending theory.

Mendel realised that he needed to carry out controlled (fair) experiments to check his ideas. He could then collect reliable evidence. Mendel took a purple-flowering plant that had produced some plants with purple flowers and some plants with white flowers. He pollinated it with a white-flowering plant and continued this over a number of plant generations. As he did this, Mendel counted the numbers of purple- and white-flowering plants that were produced.

1. What does the word 'inheritance' mean when it is used in science?

2. Which one of the following sentences A, B or C best describes the 'blending theory'.
   A A child inherits all its looks from its mother.
   B A child gets none of its looks from its parents.
   C A child's looks are a mixture of both its parents.

3. Why did Mendel disagree with the blending theory?

4. What did Mendel do to see if his ideas were right?

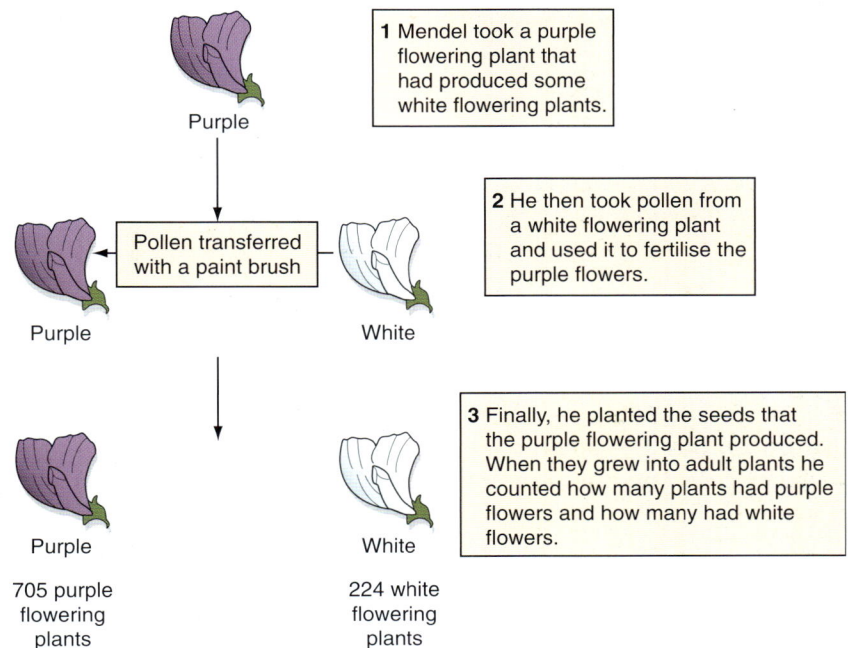

Purple

1 Mendel took a purple flowering plant that had produced some white flowering plants.

Pollen transferred with a paint brush

Purple

White

2 He then took pollen from a white flowering plant and used it to fertilise the purple flowers.

Purple

White

3 Finally, he planted the seeds that the purple flowering plant produced. When they grew into adult plants he counted how many plants had purple flowers and how many had white flowers.

705 purple flowering plants

224 white flowering plants

Figure 4.3 A flow diagram showing how Mendel investigated inheritance in pea plants

Mendel looked at his results. He noticed there were three times as many plants with purple flowers as white flowers. He concluded that some characteristics, like purple flowers, have a stronger influence in the offspring. He called these 'stronger' influences **dominant** and the 'weaker' influences **recessive**. These ideas went against the blending theory that characteristics mixed together to produce something in-between.

People ignored Mendel's results for 34 years. Then three scientists carried out some experiments similar to those of Mendel. At the same time, Mendel's work was translated into English. At last, people started to take Mendel's ideas into account. They realised he provided a far better explanation of inheritance than the blending theory.

4.2 Cell structure and inheritance

Mendel noticed how different characteristics were passed from parents to their offspring. He also showed that some characteristics were dominant and others were recessive. But Mendel couldn't explain how these characteristics were passed on. A detailed understanding of cell structure is needed to explain inheritance – the passing of characteristics from parents to their offspring.

Genes carry the information for inherited characteristics. Figure 4.4 shows how genes are small parts of a **chromosome**. There are 46 chromosomes in a normal human cell. These 46 chromosomes carry about 25 000 genes.

5 Copy and complete the following sentences. Mendel's results showed that three times as many _____ flowering plants as white flowering plants grew from his seeds. He called purple flowers the _____ characteristic. White flowers were called the _____ characteristic. There were no pink flowers, which showed that the _____ theory was not correct.

6 Write down two reasons why people ignored Mendel's results for so long.

Chromosomes are found in the nuclei of cells and are made up of genes.

A **gene** is a small part of a chromosome.

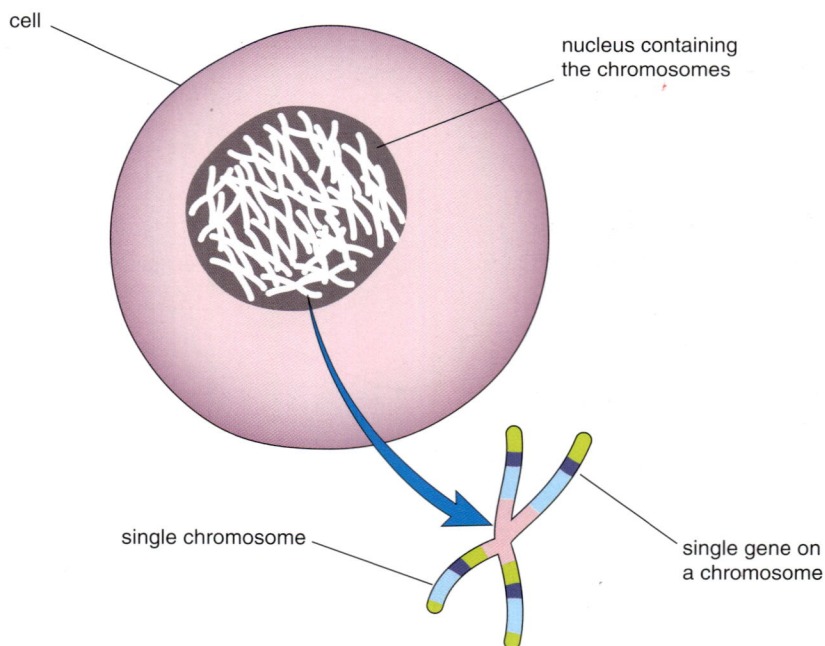

Figure 4.4 Genes are part of chromosomes in the nucleus of a cell.

cell

nucleus containing the chromosomes

single chromosome

single gene on a chromosome

Gametes are sex cells. They include sperm, egg cells and pollen.

Sexual reproduction is the joining of male and female gametes. It results in **variations** between the offspring.

---

7 Write these in order of size, largest first.

> cell   chromosome
> gene   nucleus

8 What do genes carry?

9 Why do you think humans need a large number of genes? Explain your answer.

10 Copy and complete the flow diagram below to explain how sexual reproduction produces a child.
A _____ cell from the father _____ with an egg cell from the _____.

↓

The cell that forms gets half its _____ from each parent.

↓

The cell develops into a child who inherits _____ from both parents.

Different genes control different characteristics. In Mendel's pea plants, the flower colour was controlled by one gene. Your eye colour is also controlled by one gene. These genes are passed on from parents to their children during reproduction. The male sex cells in animals are the sperm cells. In plants the male sex cells are pollen cells. Female sex cells in animals and plants are called egg cells. A general name for sex cells is **gametes**.

Sexual reproduction involves two parents. (Asexual reproduction only needs one parent. We will study this in the next section.) During sexual reproduction a new cell is formed when two gametes join (fuse) together. This new cell gets half its genes from one parent and half its genes from the other parent. The new cell develops into a new plant or animal. As the offspring inherits half its genes from its mother and the other half from its father, both parents influence its appearance and characteristics.

Children in the same family often have different looks and abilities. Scientists call these differences **variations**. Each child in a family inherits a different mixture of genes from each parent. This gives each child in a family a different set of genes unless they are identical twins. It's a bit like selecting five cards from each of two packs of cards. Every time you did this you would almost always end up with a different set of ten cards. There are about 25 000 genes in a human cell, so the combinations are almost endless.

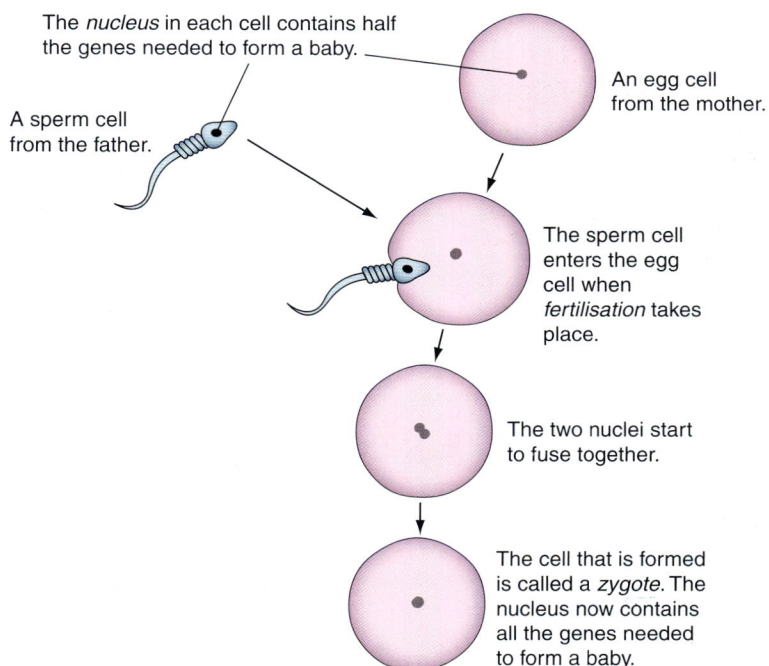

The *nucleus* in each cell contains half the genes needed to form a baby.

A sperm cell from the father.

An egg cell from the mother.

The sperm cell enters the egg cell when *fertilisation* takes place.

The two nuclei start to fuse together.

The cell that is formed is called a *zygote*. The nucleus now contains all the genes needed to form a baby.

Figure 4.5 A flow diagram showing how genes are inherited by offspring from their parents

4.3 How has our understanding of asexual reproduction helped to develop cloning?

Asexual reproduction produces offspring from only one parent. It does not involve gametes.

Cloning produces copies of an organism with the same genes as its parent.

Asexual reproduction only needs one parent. It does not involve gametes. All the genes in the offspring come from just one parent. The genes are exact copies of those in the parent. So asexual reproduction produces offspring that are genetically identical to their parent. These offspring, with genes exactly like their parent, are known as **clones**.

Cloning is not as new as you may think. Plant growers have been using cloning for a long time. Some plants, animals and micro-organisms reproduce naturally through asexual reproduction. They produce clones of themselves. Strawberry plants form stems (runners) that grow along the surface of the ground. These produce roots that grow down into the soil. The roots form new plants. The new plants have the same genes as their one parent.

⓫ Write down two differences between sexual and asexual reproduction.

⓬ What is a clone?

Bacteria are single-celled micro-organisms. They reproduce asexually. They copy all their genes and then split in half. All the copied genes go into the new cell.

Plant growers produce lots of plants cheaply using asexual reproduction. They take cuttings from older plants and these grow into new plants. Peperomia are often bought as house plants. New peperomia plants can easily be produced from older plants by taking cuttings (Figure 4.8). This is often carried out in commercial nurseries to produce hundreds of new plants.

⓭ Companies that grow and sell plants often take cuttings to produce new plants.

Decide if each of these sentences describes a benefit or a problem of taking cuttings. Write a list of the benefits and a list of the problems.
- Taking cuttings is a cheap way of producing lots of plants.
- Any problems in the original plant will be copied in all the new plants.
- People may not like the idea that the new plants are not grown from seeds.
- A very pretty plant can be copied lots of times.
- All the plants produced are very similar to look at.

Figure 4.6 Asexual reproduction of a) a strawberry plant and b) bacteria. In each case the offspring is genetically identical to its parent.

Figure 4.7 A peperomia house plant

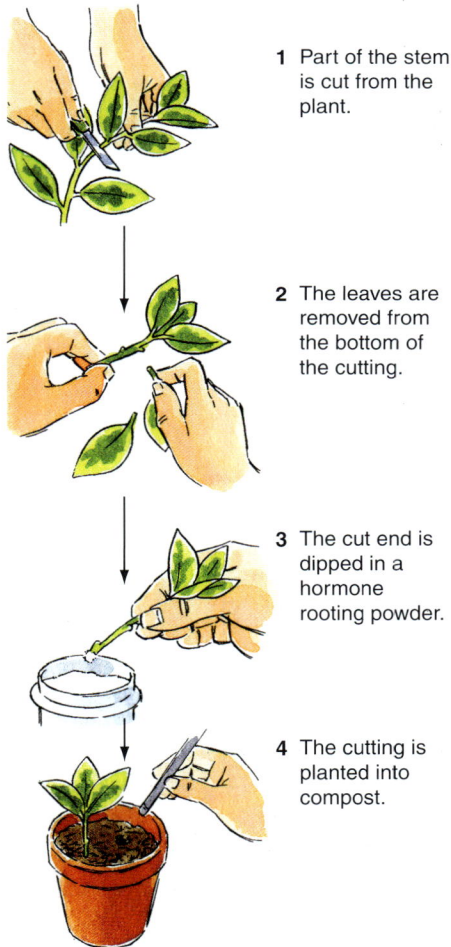

1 Part of the stem is cut from the plant.

2 The leaves are removed from the bottom of the cutting.

3 The cut end is dipped in a hormone rooting powder.

4 The cutting is planted into compost.

Figure 4.8 A flow diagram showing how a peperomia plant is grown from a cutting

Activity – Modern cloning techniques

Cloning is regularly in the news. Here are four recent headlines about cloning.

'Cloned cows may be safe to eat.'
'US scientists claim human cloning breakthrough.'
'UK court ruling means cloning is not illegal.'
'After Dolly the sheep comes Snuppy the puppy.'

It is important that you understand the science behind headlines like these. This will allow you to form your own opinion about cloning.

The information and diagrams below explain three modern cloning techniques. They will help you to think about issues related to cloning.

Tissue culture or micro-propagation is a way of producing large numbers of plants very quickly (Figure 4.9). A small number of cells are taken from a 'parent' plant. These are grown in a medium containing nutrients and plant growth hormones.

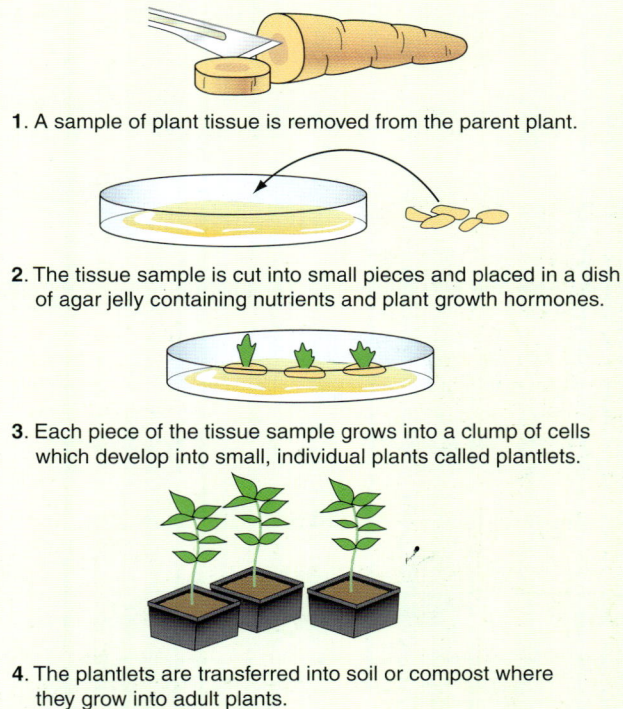

1. A sample of plant tissue is removed from the parent plant.

2. The tissue sample is cut into small pieces and placed in a dish of agar jelly containing nutrients and plant growth hormones.

3. Each piece of the tissue sample grows into a clump of cells which develop into small, individual plants called plantlets.

4. The plantlets are transferred into soil or compost where they grow into adult plants.

Figure 4.9 A flow diagram showing how tissue culture is used to produce large numbers of plants from one parent plant

Embryo transplant is a way of dividing the embryo from a pregnant animal into several parts. Each part of the embryo is then transplanted into a different host mother (Figure 4.10).

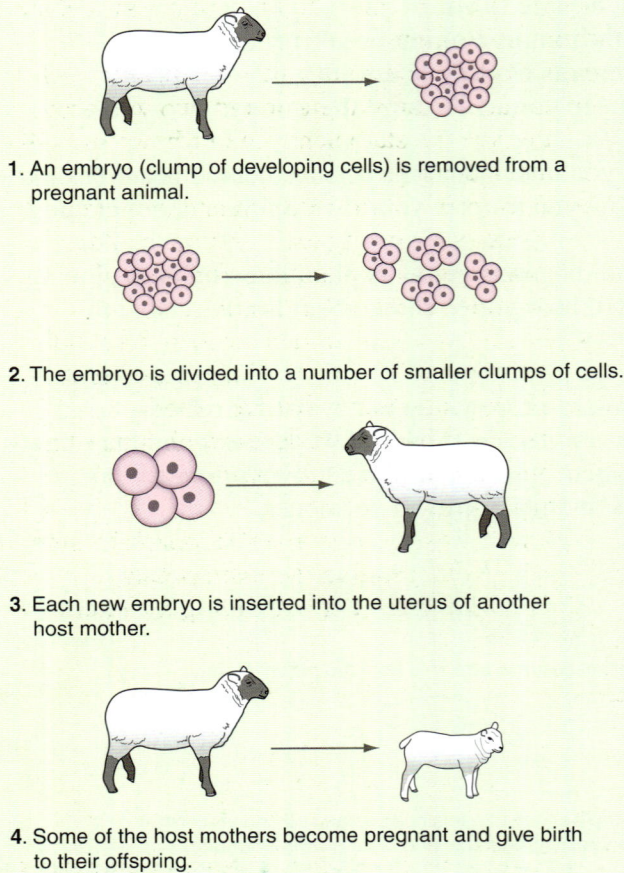

1. An embryo (clump of developing cells) is removed from a pregnant animal.

2. The embryo is divided into a number of smaller clumps of cells.

3. Each new embryo is inserted into the uterus of another host mother.

4. Some of the host mothers become pregnant and give birth to their offspring.

Figure 4.10 A flow diagram showing how embryo transplant is used to implant divided clumps of cells (new embryos) from one embryo into a number of host mothers

Fusion cell and adult cloning is used to produce exact copies of cells or whole individuals from a single cell (Figure 4.11).

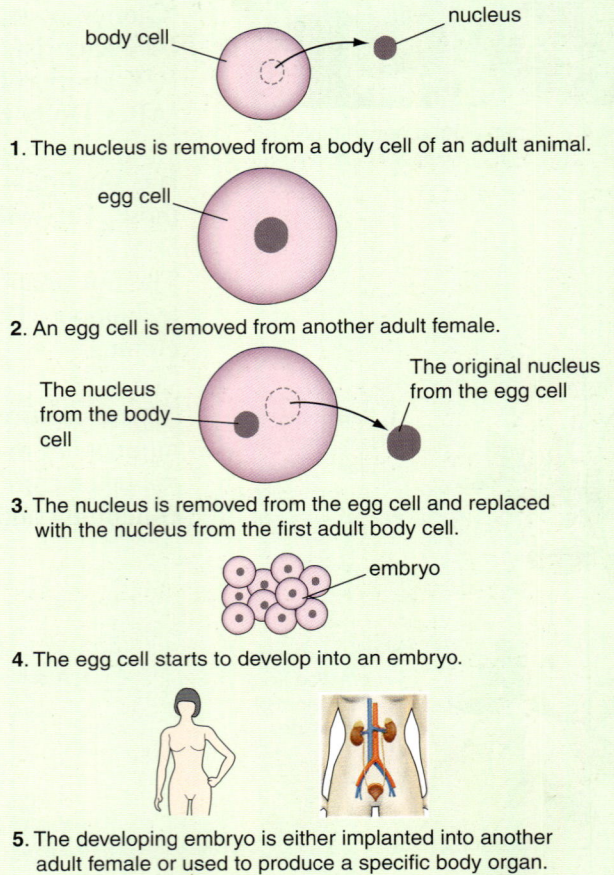

body cell   nucleus

1. The nucleus is removed from a body cell of an adult animal.

egg cell

2. An egg cell is removed from another adult female.

The nucleus from the body cell

The original nucleus from the egg cell

3. The nucleus is removed from the egg cell and replaced with the nucleus from the first adult body cell.

embryo

4. The egg cell starts to develop into an embryo.

5. The developing embryo is either implanted into another adult female or used to produce a specific body organ.

Figure 4.11 A flow diagram showing how fusion cell and adult cloning is used to produce cloned animals and human body organs

❶ Copy and complete Table 4.1 showing what each cloning technique is used for.

Type of cloning	What is the cloning technique used for?
Tissue culture	
Embryo transplant	
Fusion cell and adult cloning	

Table 4.1 The uses of different cloning techniques

Benefits of tissue culture	Drawbacks of tissue culture
A lot of new plants can be grown in a short time	All plants have the same genes, so they may all suffer from the same diseases or pests
Little space is needed, and conditions can be precisely controlled	New beneficial characteristics cannot arise by mixing genes from different plants
All new plants inherit the same characteristics	There is no variation in the plants. This increases the danger of reducing the gene pool

Table 4.2 Three benefits and three drawbacks of using tissue culture to clone plants

2 Look carefully at Table 4.2. This shows three benefits and three drawbacks of using tissue culture.

3 Discuss in pairs the benefits and drawbacks of using embryo transplant techniques.

4 Draw a table for embryo transplant techniques showing two benefits and two drawbacks.

5 a) Discuss in pairs the benefits and drawbacks of using fusion cell and adult cloning techniques.

b) Draw a table for fusion cell and adult cloning techniques showing two benefits and two drawbacks.

6 'Research into cloning humans is wrong and should be banned.' This is a view held by many people. You are going to prepare a mini-debate about this viewpoint.

- Get into a group of five. Two people will argue *for* the statement and two will argue *against* the statement. The fifth person will act as a judge.
- Each pair should research and prepare a speech supporting their argument. The judge should research the topic and become an expert on cloning.
- Each pair will then deliver their speech to the other pair while the judge takes notes of the main points that are raised.
- The judge will then decide which pair has put forward the most convincing argument.
- Finally, each judge gives feedback to the whole class stating the main points raised by each pair and giving his / her verdict.

4.4 Understanding genetic engineering

Figure 4.12 A diabetic injecting herself with genetically engineered insulin. In the past, insulin had to be taken from slaughtered animals.

Figure 4.13 People protesting against genetically modified crops

Research on genetic engineering has been going on for more than 40 years. Genetic engineering could have many useful applications, but some people are seriously concerned about its use. It is important that you understand what genetic engineering involves and how it can be used. This will help you to make up your mind about its benefits and its dangers.

Genetic engineering of the human growth hormone

Some children suffer from a condition called pituitary dwarfism. They grow very slowly and reach puberty much later than usual. This happens because their pituitary gland does not make enough growth hormone. At one time, these children were given extra growth hormone taken from dead people. But, this was risky. Children treated with the growth hormone could develop a disease of the nervous system called CJD. The use of human growth hormone was stopped when this risk was discovered.

Human growth hormone is now produced by genetic engineering. The gene that produces the growth hormone is taken from a healthy person. This gene is then put into a bacterial cell. The bacterial cell reproduces asexually making clones of itself. All the new bacteria contain the human growth hormone gene. So they all produce human growth hormone. This is collected and used to treat children suffering from pituitary dwarfism. The treatment is expensive but very effective.

1. The gene that produces growth hormone is removed from the nucleus of the cell.

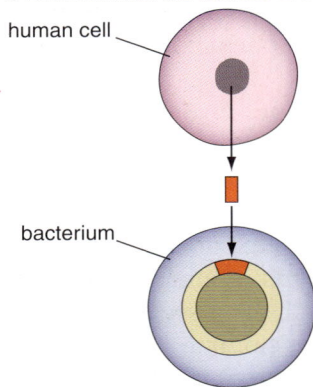

human cell

bacterium

2. The gene is then inserted into the DNA of a bacterium. The bacterium then starts producing human growth hormone.

Figure 4.14 A flow diagram showing how genetically engineered human growth hormone is produced

⑭ Read the three sentences below. Write down the one that best describes genetic engineering.
  • Genetic engineering means making clones.
  • Genetic engineering involves taking a gene from one cell and putting it into another cell.
  • Genetic engineering changes all the chromosomes in a living thing.

⑮ What is the benefit of using genetically engineered human growth hormone instead of growth hormone taken from dead people?

⑯ Table 4.3 lists some uses of genetic engineering. Copy and complete the table stating whether you are for or against each use. Give one reason for each decision. The first one has been done as an example.

Uses of genetic engineering	For or against?	Reason
Producing human insulin from genetically modified bacteria	For	Animals no longer have to be killed to provide insulin for diabetics
Producing disease-resistant rice to increase yields in poor countries		
Producing chickens with four legs and no wings to increase meat production		
Parents being able to choose the sex of their baby		
Developing plants that produce plastics, reducing the need for crude oil		

Table 4.3 Uses of genetic engineering

4.5 How do new species of plants and animals evolve?

There is an amazing variety of life on Earth. Scientists estimate that there may be 50 million different types of plants and animals. How did all these different animals and plants come to be living on Earth? How have these species developed from their ancestors? Although scientists can answer some questions like these, they are still not sure how life began on Earth.

The plants and animals that we see today have developed from ancestors that lived in the past. These ancestors were different from today's species. Scientists say that the present plants and animals have **evolved** (gradually changed) from their ancestors. They call this process of gradual change **evolution.** Evolution has resulted in species that are adapted to survive in the environment in which they live. No one has actually seen these changes taking place because the process takes so long. But there have been several theories that try to explain evolution.

Some theories of evolution

Four different theories for evolution are described below. Read them and think about the ways in which the theories disagree. Then try to decide which one offers the best explanation of evolution.

1 Creationism

Creationism suggests that every species was created separately by God. It says that species do not change through time. This was the view once held by the Christian Church. It was strongly supported by Archbishop James Ussher in the seventeenth century. He worked out from the Bible that all species alive today were created about 6000 years ago.

2 Buffon's theory

George Leclere Buffon, a French biologist, put forward the second theory. He suggested that the Earth was at least 75 000 years old. Buffon thought that certain species changed through time. He explained that these changes could be caused by the environment. Sometimes they happened by chance. He even suggested that humans and apes were related. Buffon kept his ideas quiet and did not publish them widely.

3 Lamarck's theory

Jean Baptiste Lamarck put forward the third theory. He believed that an animal could change gradually in its own lifetime. He also said that these changes were then passed on to its offspring. For example, he explained how herons had evolved long legs. Lamarck said that herons in the past had stretched their legs to stay dry when wading in water. These herons then passed on this characteristic to their offspring. Lamarck also used this theory to explain why giraffes have such long necks.

Figure 4.15 Some of the animals and plants found on the Earth today

Figure 4.16 Archbishop James Ussher (1581–1656) believed strongly in creationism.

Figure 4.17 George Leclerc Buffon (1707–1788) was the first scientist to suggest that species might evolve over thousands of years.

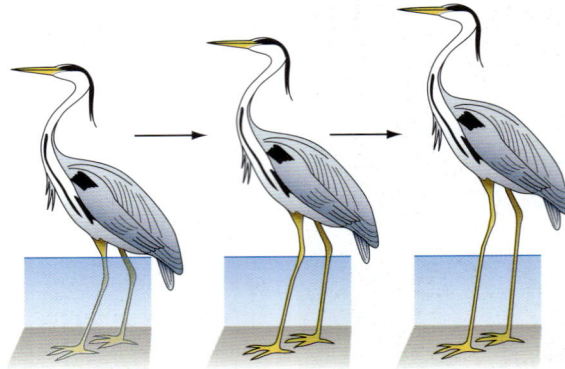

Figure 4.18 Lamarck's explanation for the heron's long legs

4 Darwin's theory

The final theory to consider was put forward by Charles Darwin in the nineteenth century. Darwin studied the animals and plants on the Galapagos Islands. These islands are 600 miles from the west coast of South America. Darwin found that the islands had species of animals that were not found anywhere else in the world. For example, he found 13 different species of finch on the islands. Only one of these species lived on the mainland of South America. The different finch species on the Galapagos Islands had differently shaped beaks. Their beaks suited the food they ate. For example, one species had a very strong beak for cracking nuts. Another species, living on a different island, had a long thin beak for picking up insects.

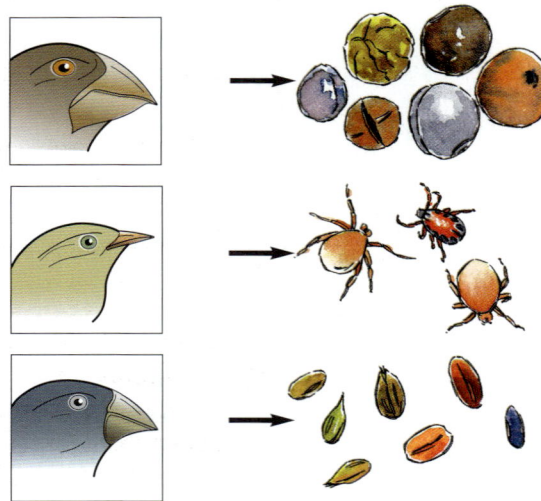

Figure 4.19 Three types of finch found on the Galapagos Islands and the food they eat

Darwin suggested that the different finches on the Galapagos Islands had evolved from a single species. He said that within a population of finches there were always variations in features, such as beak size and shape. Individual birds were more likely to survive if they had features that suited their environment better. These surviving birds would pass on these features to their offspring. Darwin found a similar pattern in other species he studied on the Galapagos Islands.

Figure 4.20 This drawing of Darwin's head on an ape's body was printed after he published his theory of evolution.

Activity – Analysing theories of evolution

❶ Copy and complete Table 4.4 summarising the main points in each theory. The first one has been done for you.

Theory or scientist	Main points
Creationism	• God created all species alive today about 6000 years ago • Species have not changed over time
Buffon	
Lamarck	
Darwin	

Table 4.4 Summary of the theories for evolution

❷ Why do you think Buffon kept his ideas quiet and did not publish them?

❸ Copy and complete the sentences below using the words in this box.

animal	genes	lifetime	offspring	possible

Lamarck said that changes to an _____ in its lifetime would be passed on to its _____. We now know that this is not _____ because an animal's characteristics are controlled by its _____ and these do not change during its _____.

❹ Darwin's theory of evolution is now widely accepted. Write down two differences between Darwin's theory and the other theories.

❺ Darwin's ideas were only gradually accepted. Why do you think that a cartoon like that in Figure 4.20 was printed?

❻ Creationism reflected a religious belief about life on Earth. Today Christians still believe God created life. Many of them also accept Darwin's theory of evolution. Discuss in pairs how Christians can believe that God created life on Earth and also accept Darwin's theory of evolution.

4.6 Evidence for evolution and natural selection

Darwin based his theory of evolution on observations he made on the Galapagos Islands and around the world. Today we have much more evidence to support Darwin's theory. The theory of evolution says that all animals and plants have evolved from simple life forms like bacteria. These appeared on the Earth three thousand million (3 billion) years ago. Humans only appeared about one hundred thousand years ago. But, how do we know what the forerunners of today's species were like? How did we get our evidence for evolution?

Evidence for evolution comes from fossils. Scientists study fossils. They find out what the animals and plants of the past looked like. Fossils show how animals and plants have changed. Fossils of crocodile jaws and teeth show that they have changed very little over the last 200 million years (Figure 4.21). The earliest fossils of our human ancestors such as *Orrorin tugenensis*, which was the size of a chimpanzee, show that humans have evolved a great deal in 2 million years.

Fossils show how whales could have evolved from a mammal that lived on land. Look at Figure 4.22. This shows how fossil records have been used to work out what the ancestors of whales may have looked like.

These pictures show how the ancestors of whales gradually changed. Slowly, they became better suited to living in water. Those better suited to living in water would be most likely to survive. They would be more likely to reproduce and their offspring would inherit these features. The best swimmers among their offspring would also survive and breed most successfully, and so on.

Figure 4.21 Fossils of a crocodile's jaw and teeth from 200 million years ago (the larger) and those of a modern-day crocodile (the smaller)

17 What can scientists find out from studying fossils?

18 Why are the fossils of a whole skeleton more useful than only part of it?

19 Copy and complete the sentences below using the words in this box.

> adapted   evolution
> generations   genes
> inherited   land   selection

Whales may have evolved from a mammal that lived on _____. Natural _____ gradually led to changes that made it more _____ to living in water. The _____ that controlled the characteristics making it suited to living in water were _____ by its offspring This happened over many _____ of the species, eventually leading to the _____ of the modern whale.

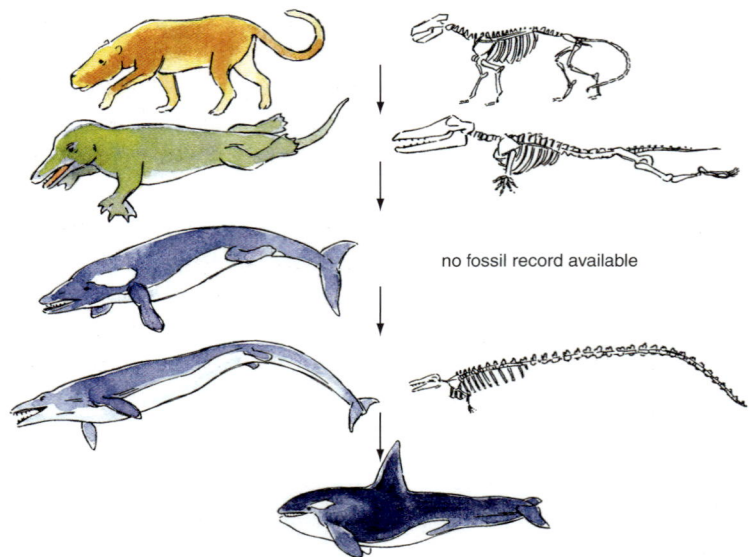

no fossil record available

Figure 4.22 How whales may have evolved from a mammal living on land

This sequence of events that leads to evolution is called **natural selection**. Natural selection can be divided into five key steps.

1   The individuals within a species vary because of differences in their genes. These variations arise because individuals inherit different genes or because genes occasionally mutate (change).
2   Individuals with the characteristics that best suit their environment are more likely to survive.
3   These individuals are also more likely to reproduce successfully.
4   The genes which have enabled these individuals to survive are then passed on to their offspring.
5   This process is repeated over many generations, leading eventually to a new species.

4.7 Why do species become extinct?

An animal or plant is extinct when there are no more living members of its species. Extinction is often part of evolution. As a new species evolves, the old one dies out. Sometimes a species becomes extinct without a new species evolving. The main causes of extinction are:
*   changes in the environment;
*   new predators (animals that eat it);
*   new diseases;
*   new competitors.

It can be difficult to work out the cause of extinction. For example, a bird called a dodo became extinct about 300 years ago. The dodo was a flightless bird that lived on the island of Mauritius. It finally became extinct after humans had settled on the island. For many years, people thought the dodo became extinct because it was hunted. In other words, humans were a new predator to the dodo. But scientists have now discovered that the dodo was not often eaten by the people on the island.

The humans who moved to Mauritius cut down most of the forests to make space for their homes. This destroyed the dodo's habitat. So changes in the environment may have caused the extinction of the dodo.

The settlers brought animals with them to Mauritius. Most of these were new to the island. They included monkeys, pigs and rats. These animals ate the same food as the dodos. They also destroyed the dodo's nesting sites. The new animals competed with the dodos for habitat and food. So these new competitors may have caused the extinction of the dodo.

Figure 4.23 The dodo became extinct around 1700.

20 Scientists think that the modern horse evolved from an ancestor the size of a fox. Write the following sentences in the correct order to explain how the horse may have evolved by natural selection.
- Gradually, over many generations, taller horses with toes joined together were produced.
- Some *Eohippus* had slightly longer legs than others and could run faster.
- Some *Eohippus* also had toes that were closer together which helped them to run even faster.
- Eventually the modern horse evolved, with long legs and hooves instead of toes.
- The individuals that survived were more likely to reproduce and pass on their characteristics to their offspring.
- Sixty million years ago, a small mammal, *Eohippus*, existed that had three toes and short legs.
- These individuals were better at avoiding predators and were more likely to survive.

21 Why are we still not sure what caused the dodo to become extinct?

22 Humans have caused a large increase in the number of species that have become extinct.
  a) Which human activities do you think have caused this increase?
  b) What problems could this loss of species cause in the future?

Summary

✓ Some characteristics are passed on from parents to their offspring. This is called **inheritance**.

✓ The information that controls inherited characteristics is carried by **genes**. Genes are passed on in the sex cells (**gametes**).

✓ A gene is a section of a **chromosome**. The chromosomes are in the nucleus of a cell.

✓ **Sexual reproduction** is the joining of a male and a female gamete. Sexual reproduction mixes genetic information from two parents. This leads to variation in their offspring.

✓ **Asexual reproduction** only needs one parent and does not involve the joining of gametes. Offspring produced by asexual reproduction from one parent are genetically identical and show no variation.

✓ Genetically identical individuals are called **clones**.

✓ New plants can be produced quickly and cheaply by taking cuttings. This is a type of cloning.

✓ Modern cloning techniques include tissue culture, embryo transplants, fusion cell and adult cell cloning.

✓ It is important to be able to make informed judgements about the economic, social and ethical issues in the use of cloning.

✓ **Genetic engineering** can be used to transfer genes from the cells of one organism to the cells of other organisms.

✓ Genes can be transferred to the cells of plants or animals to introduce new characteristics to the organism.

✓ It is important to be able to make informed judgements about the economic, social and ethical issues of genetic engineering.

✓ There have been a number of different theories to explain **evolution**.

✓ Darwin's theory of evolution is now the most widely accepted. Darwin's theory explains how evolution occurs by **natural selection**.

✓ The theory of evolution states that all species of living things have evolved from simple life forms. These first appeared more than 3 billion years ago.

✓ Fossils provide us with strong evidence to support the theory of evolution.

✓ The **extinction** of a species may be caused by changes in the environment, new predators, new diseases or new competitors.

✓ It is not always easy to decide what caused a particular species to become extinct.

1 Young rabbits, like these, often look like their parents.

This is because information about their appearance, for example fur tone, is handed down from parents to their offspring.

Copy and complete the paragraph below using the appropriate words from the box.

> body   chromosomes   clones
> cytoplasm   genes   nucleus   sex

Information is passed from parents to their offspring in _____ cells. Appearance is controlled by _____. The structures which contain information about a large number of characteristics are known as the _____. In the cell these structures are found in the _____.

(4 marks)

2 Humans reproduce by sexual reproduction. Michael and James are brothers, but they have very different features.

ZOOT

a) State two ways in which sexual reproduction is different from asexual reproduction. *(2 marks)*
b) Explain why Michael and James have different features. *(3 marks)*

3 Read the piece of text below carefully.

The first cloned pet in the United States is named Little Nicky, a 9-week-old kitten. It belongs to a Texas woman saddened by the loss of a cat she owned for 17 years. The kitten cost its owner $50 000 and was created from DNA from her beloved cat, named Nicky, who died last year. 'He is identical. His personality is the same,' the owner, Julie, told The Associated Press in an interview. She asked that her surname and hometown were not disclosed, because she was afraid of being targeted by groups opposed to cloning.

a) Explain why Little Nicky is exactly the same as Nicky. *(3 marks)*
b) Little Nicky's owner is concerned that she may become a target for anti-cloning groups
   i) Suggest **one** group of people who might be opposed to the cloning of pet cats. *(1 mark)*
   ii) Explain why this group might be opposed to cloning. *(1 mark)*

4 a) Provide an explanation for the theory of evolution. *(2 marks)*
b) The passage below provides a possible explanation for the evolution of the long legs of wading birds, by Lamarck.

When an animal undergoes change during its lifetime and then mates, it passes this change on to its offspring. The ancestors of wading birds needed to feed on fish. In trying to get into deeper water, while keeping their bodies dry, they stretched their legs as much as they could. In doing this again and again they made their legs slightly longer. This trait was passed on to the next generation, who in turn stretched their legs. Over many generations, the wading birds' legs became much longer.

Darwin would have provided a different explanation for the evolution of the long legs of these birds. What are the main differences between his explanation and that of Lamarck? *(3 marks)*

Chapter 5
How do rocks provide useful materials?

At the end of this chapter you should:

✓ understand that atoms are the smallest particles in elements;

✓ understand that atoms can join together to form molecules in compounds;

✓ be able to use symbols and formulae to write balanced equations for chemical reactions;

✓ know that rocks provide stone for building, metals and alloys;

✓ know how limestone is used to manufacture quicklime, slaked lime, cement, concrete and glass;

✓ understand how metals can be extracted from their ores;

✓ have considered the social, economic and environmental effects of quarrying, mining and extracting metals;

✓ have considered the benefits and drawbacks of using metals and recycling metals.

Figure 5.1 Iron ore is being mined in this photograph. Iron ore like this is useless. You can't grow anything in it, you can't eat it and you can't build with it. But, if it is heated with limestone, coke and air, it produces iron. From iron we can make steel, which is one of our most useful materials.

5.1 What sorts of materials are there?

Materials that occur naturally, like iron ore, limestone, water and air are called **raw materials** or **naturally occurring materials**. Materials like iron and steel don't occur naturally. But we can make them from raw materials such as iron ore. Because of this, iron and steel can be called **manufactured materials**. These manufactured materials are useful products, which we need for everyday life.

What are our most important raw materials?

Table 5.1 shows the five most important raw materials and some of the manufactured materials we get from them.

The manufactured materials are:
- either separated from the raw materials by processes such as distillation;
- or made from the raw materials by chemical reactions such as iron from iron ore.

Figure 5.2 At one time, football boots were made of leather. Today, football boots are made from polymers such as PVC and polyester.

❶ What raw material does leather come from?

❷ What raw material are PVC and polyester made from?

❸ a) Write down two benefits of using glass for the building in Figure 5.3.
b) Write down two drawbacks and risks of using glass.

Figure 5.3 The outside of this building is mainly glass.

Raw material	Manufactured materials obtained from the raw material
Rocks	Metals (iron, aluminium, copper) Alloys (steel, brass) Building materials (cement, glass)
Crude oil	Fuels (petrol, diesel) Plastics (polythene, PVC and polyester)
Air	Nitrogen for making ammonia, nitric acid and fertilisers Oxygen for breathing equipment
Sea water	Table salt, sodium hydroxide, chlorine and hydrogen from brine (concentrated sodium chloride)
Plants	Fruit and vegetables Plant oils for cooking and medicines Fuels from biomass materials

Table 5.1 The five most important raw materials

❹ Copy and complete the following table. The first line has been done for you.

Manufactured material	Which raw material in Table 5.1 did the manufactured material come from?
Metals in your TV Plastics in your MP3 player Ammonia in cleaning fluid Olive oil for cooking Polyester in your shirt/blouse Glass for bottles and jars	Rocks

The properties of a material show how it differs from other materials. They also help us to put similar materials into groups or sets. For example, in this chapter we have put materials into two groups – **naturally occurring** and **manufactured**. When we put similar things into groups, we usually say that we classify them.

During KS3, it was useful to classify materials as **solids**, **liquids** and **gases** – the **three states of matter**.

Another useful way to classify materials is as **elements**, **mixtures** and **compounds**. We will look at this in Section 5.2.

Activity – Using the natural resources on Geecee

Read the following information about the island of Geecee. Look at Figure 5.4. Then answer the questions.

Geecee is an imaginary island off the west coast of Britain. The island has strong winds blowing from the west. There are high mountains with fast flowing rivers. There are thick forests along the west coast and peat bogs in the south-east. There are no supplies of coal, oil or natural gas. At present no one lives on Geecee.

You have been asked to study whether people could live on the island and start a fishing industry.

1. What naturally occurring material would you use to build the first homes on Geecee? Explain your answer.
2. Describe two ways of producing heat to cook food on the island.
3. Where would you build a port for the fishing boats? Give a reason for your choice of site.
4. Where would you build homes on the island? Explain your answer.
5. Name one naturally occurring material that you would try to conserve (keep). Explain your answer.

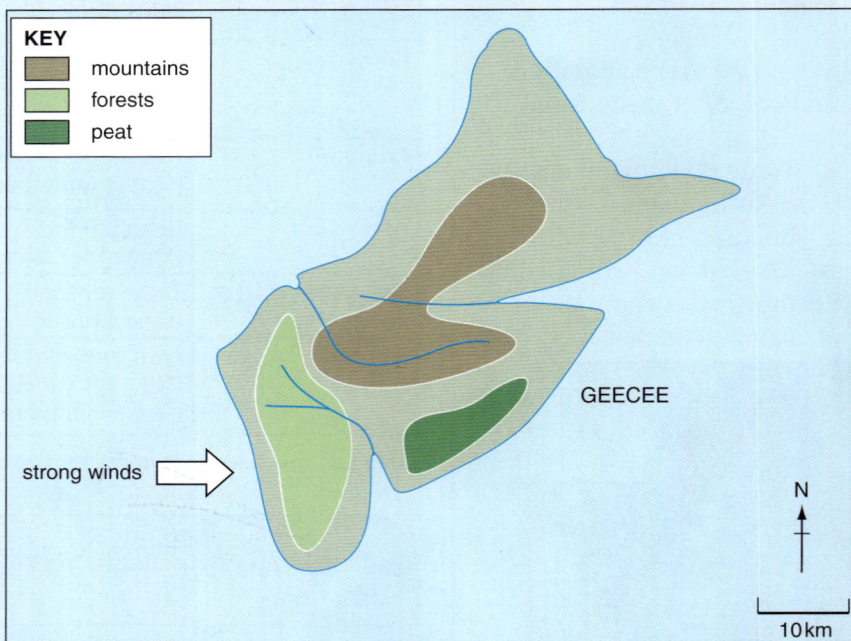

Figure 5.4 Map of Geecee

Elements, compounds and mixtures

When electricity is passed through molten sodium chloride, it breaks down to form sodium and chlorine (Figure 5.5). We can summarise the **chemical reaction** by writing a **word equation**.

sodium chloride → sodium + chlorine

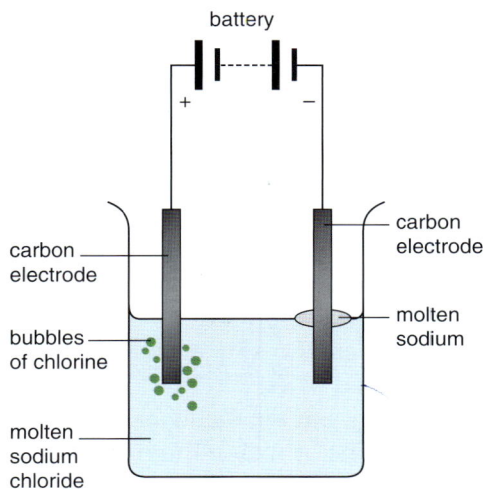

Figure 5.5 When electricity is passed through molten sodium chloride, it breaks down into sodium and chlorine.

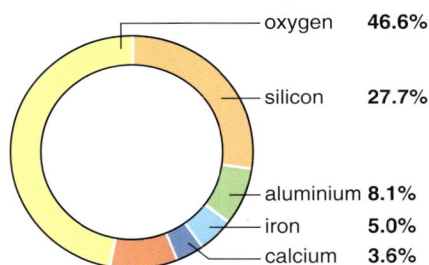

Figure 5.6 The percentages of the commonest five elements in the Earth's crust

A **mixture** is two or more substances which are mixed together. In a mixture, the substances are *not* combined chemically.

No matter what you do to the sodium and chlorine, they can't be broken down into simpler substances.

Substances that cannot be broken down into simpler substances are called **elements**. Elements are the simplest possible materials. Aluminium, iron, oxygen and carbon are examples of other elements. Substances like sodium chloride and water are not elements because they can be broken down into simpler substances. Substances like sodium chloride and water that contain two or more elements chemically joined together are called **compounds**.

There are about 100 different elements. Although there are millions and millions of different substances in the universe, they all contain one or more of these elements. For example, water contains hydrogen and oxygen. Sand is made of silicon and oxygen. Limestone contains calcium, carbon and oxygen. So elements are the simplest building blocks for all substances.

Figure 5.6 shows the percentages of the commonest five elements in the Earth's crust.

5 a) What total percentage of the Earth's crust do these five elements make up?
b) What percentage of the Earth's crust do all the other elements make up?
c) Which metal forms the highest percentage in the Earth's crust?
d) A student used five different textbooks to check the percentage of sodium in the Earth's crust.
The values given were 2.7%, 3.5%, 2.6%, 2.9% and 3.0%.
Copy and complete the following sentences.
These values vary from the minimum value of _____% to the maximum value of _____%. This is called the **range in the data**. One of the values given is very different from the other four. This **anomalous** value is _____%. Anomalous means unusual or unexpected. These unusual values are usually ignored in calculating an average or mean value. When the anomalous value is ignored:
The average (mean) value for the percentage of sodium in the Earth's crust = _____%.

Most of the materials that occur naturally and that we use every day are *not* pure elements or pure compounds. They are **mixtures** of substances. They may be:
- mixtures of elements, such as **alloys**, which are metals mixed with other elements (for example, mild steel is mainly iron with about 0.2% carbon);
- mixtures of compounds, such as seawater, which contains salt (sodium chloride) and water.

Figure 5.7 This photo of a gold crystal was taken through an electron microscope. Each yellow blob is a separate gold atom. The magnification of the photo is 40 million.

What are the particles in elements?

The smallest particles of an element are **atoms**.

Electron microscopes can magnify objects more than a million times. By using an electron microscope, we can see atoms. Figure 5.7 shows an electron microscope photo of gold.

> **6** a) Measure the diameter of one gold atom in the photo.
> b) Now copy and complete the following sentences by choosing the correct word in each of the underlined pairs.
> A more accurate/successful way to work out the diameter of a gold atom in the photo is to measure a line of five of them and then divide by 40 million/5. This gives an average/equal value which is more reliable/useful because any change/error in your measurement is divided by 5.
> To work out the actual diameter of a gold atom you divide/multiply the diameter on the photo by 40 million.

All substances are made of elements and all elements are made of atoms. Therefore all substances are made of atoms.

Each element contains only one sort of atom. So, as there are about 100 different elements, there are also about 100 different kinds of atom. But only iron contains iron atoms, only copper contains copper atoms, and so on.

> **7** Why is it useful to represent elements using symbols? (Hint: What would you do if you had to write 'magnesium' many times?)
>
> **8** Use the Periodic Table containing symbols on page 83 to answer the following questions.
> a) What are the symbols for carbon, calcium, cobalt, copper, chlorine and chromium?
> b) What elements are represented by N, Ni, P, K, Si, Ag, S and Na? Write out a list with the symbol followed by the element's name.

Representing atoms with symbols

Atoms of each element can be represented by a chemical **symbol**. For example, O represents an atom of oxygen, Fe represents an atom of iron and C represents an atom of carbon. The symbol for an element can also be used as shorthand for the element itself. So O means oxygen, Fe means iron, and so on.

The names and symbols of all common elements are shown in the Periodic Table on page 83.

Sometimes it is useful to draw pictures of atoms as coloured circles with the symbol in the centre (Figure 5.8). These coloured circles can help us to understand the structure and reactions of substances.

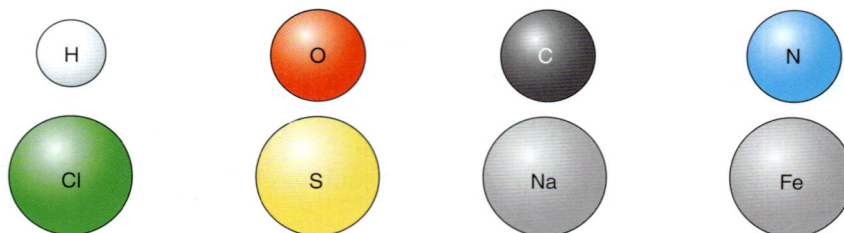

Figure 5.8 Pictures of atoms. Hydrogen atoms are usually shown as white circles, oxygen red, carbon black, nitrogen blue, chlorine green, sulfur yellow and metals grey.

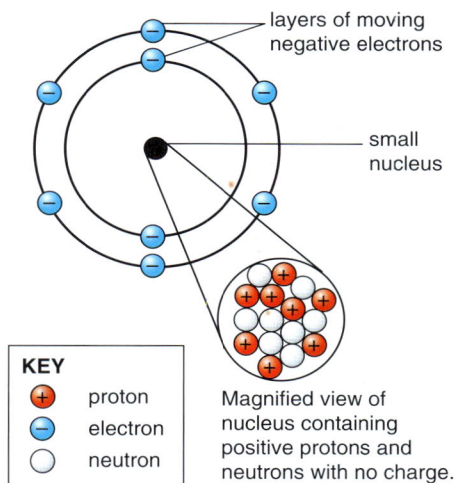

KEY
- ⊕ proton
- ⊖ electron
- ○ neutron

Magnified view of nucleus containing positive protons and neutrons with no charge.

Figure 5.9 Protons, neutrons and electrons in an atom of oxygen

A **molecule** is a particle containing two or more atoms joined by chemical bonds.

An **ion** is a charged particle formed when an atom loses or gains one or more electrons.

How do atoms combine?

Atoms have a small central **nucleus**. Around the nucleus, there are negative **electrons**. The nucleus contains two kinds of particle: **protons** and **neutrons**. Protons are positive and neutrons are neutral (Figure 5.9). The positive charge on one proton cancels the negative charge on one electron. Atoms have the same number of protons as electrons. So the overall charge on an atom is neutral.

The number of protons in an atom is its **atomic number**. So hydrogen atoms with one proton have an atomic number of 1. Helium atoms with two protons have an atomic number of 2, oxygen atoms with eight protons have an atomic number of 8, and so on. There is more about atomic numbers in Section 11.7.

When elements react, their atoms join with other atoms to form compounds. This involves:
- either sharing electrons to form **molecules**;
- or transferring (giving and taking) electrons to form charged particles called **ions**.

When atoms of non-metals join together, they share electrons and form molecules.

In a molecule of water, two atoms of hydrogen are combined with one atom of oxygen. The two hydrogen atoms and the one oxygen atom are held together by strong **chemical bonds** (Figure 5.10). Where the outer parts of the atoms overlap, electrons are shared. These shared electrons are attracted by the nuclei of both atoms and this holds the atoms together.

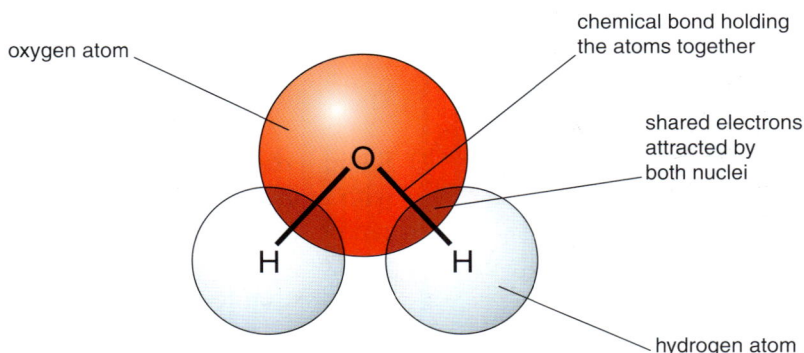

Figure 5.10 A molecule of water showing bonds and shared electrons

The symbols for elements are also used to represent molecules in compounds. So water molecules, which have two hydrogen atoms (H) and one oxygen atom (O), are written as H_2O. Carbon dioxide is written as CO_2 – one carbon atom (C) and two oxygen atoms (O).

H_2O and CO_2 are called **molecular formulae**, or just formulae for short. The word 'formulae' is the plural of the word 'formula'.

Molecular formulae show the number of atoms of the different elements joined together in one molecule of a compound.

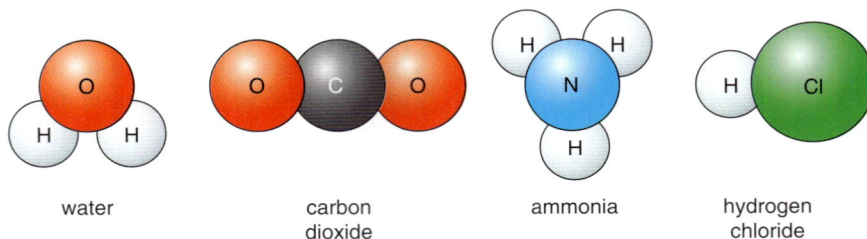

water   carbon dioxide   ammonia   hydrogen chloride

Figure 5.11 Pictures of the molecules of water, carbon dioxide, ammonia and hydrogen chloride

9 Look at Figure 5.11.
  a) In one molecule of ammonia, how many atoms are there of i) nitrogen, ii) hydrogen?
  b) In one molecule of hydrogen chloride, how many atoms are there of i) hydrogen, ii) chlorine?

10 What is the formula for i) ammonia, ii) hydrogen chloride?

11 a) Draw a picture for a molecule of methane (natural gas), CH_4. All the four hydrogen atoms are bonded to the carbon atom.

  b) Copy and complete the sentences below using the words in this box.

atom	CH_4	five	four
hydrogen	molecule		two

The formula for methane is _____. In methane, one _____ of carbon has joined with _____ atoms of _____ to form a _____ of methane. One molecule of methane contains _____ different elements and a total of _____ atoms.

When atoms of a metal join with atoms of a non-metal, they form ions.

Sodium chloride (NaCl) forms when atoms of sodium (a metal) join with atoms of chlorine (a non-metal). When sodium chloride forms, each sodium atom gives up one electron to form a sodium ion (Na^+) with one positive charge. At the same time, each chlorine atom takes one electron to form a chloride ion (Cl^-) with one negative charge.

In solid sodium chloride, these ions form a lattice structure (Figure 5.12). The opposite charges on the ions attract each other. This forms strong chemical bonds between the ions. These chemical bonds hold the ions together in the lattice.

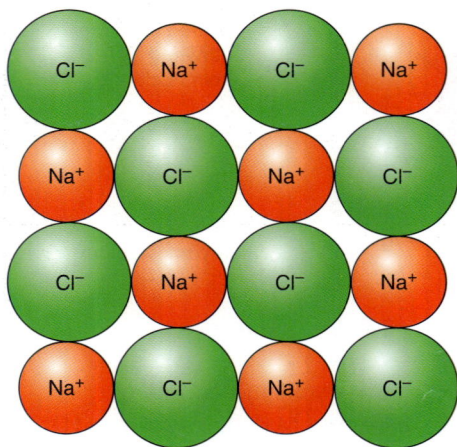

Figure 5.12 The lattice of Na^+ and Cl^- ions in one layer of solid sodium chloride

Calcium oxide (CaO) and red iron oxide (Fe_2O_3) are also formed when a metal reacts with a non-metal. These substances also have lattices with ions.

an oxygen atom

Symbol O

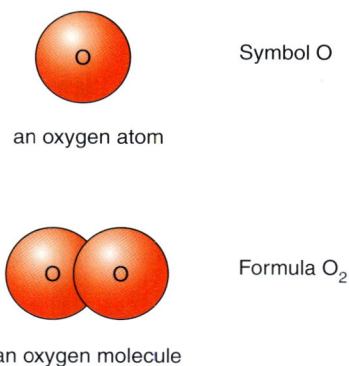
an oxygen molecule

Formula O_2

Atoms and molecules of elements

Almost all elements can be represented by their symbols, for iron, C for carbon, and so on. But this is not the case oxygen, nitrogen and chlorine. These elements exist as mo containing two atoms joined together. So hydrogen is best represented as H_2 not H, oxygen as O_2 not O, nitrogen as N_2 and chlorine as Cl_2 (Figure 5.13).

Figure 5.13 An atom and a molecule of oxygen

5.3

How are elements arranged in the Periodic Table?

In the **Periodic Table** (Figure 5.14), elements are arranged in order of their atomic number. (Remember that the atomic number is the number of protons in one atom of an element.) So the first element in the Table is hydrogen (atomic number 1), then helium (atomic number 2), then lithium (atomic number 3), and so on.

Group	1 Alkali metals	2 Alkaline -earth metals											3	4	5	6	7 Halogens	0 Noble gases
Period																		
1				Key														He helium 2
2	Li lithium 3	Be beryllium 4											B boron 5	C carbon 6	N nitrogen 7	O oxygen 8	F fluorine 9	Ne neon 10
3	Na sodium 11	Mg magnesium 12			transition elements								Al aluminium 13	Si silicon 14	P phosphorus 15	S sulfur 16	Cl chlorine 17	Ar argon 18
4	K potassium 19	Ca calcium 20	Sc 21	Ti 22	V 23	Cr chromium 24	Mn manganese 25	Fe iron 26	Co cobalt 27	Ni nickel 28	Cu copper 29	Zn zinc 30	Ga 31	Ge 32	As 33	Se 34	Br bromine 35	Kr krypton 36
5	Rb 37	Sr 38	Y 39	Zr 40	Nb 41	Mo 42	Tc 43	Ru 44	Rh 45	Pd 46	Ag silver 47	Cd 48	In 49	Sn tin 50	Sb 51	Te 52	I iodine 53	Xe 54
6	Cs 55	Ba 56	La 57	Hf 72	Ta 73	W 74	Re 75	Os 76	Ir 77	Pt platinum 78	Au gold 79	Hg mercury 80	Tl 81	Pb lead 82	Bi 83	Po 84	At 85	Rn 86
7	Fr 87	Ra 88	Ac 89	Rf 104	Db 105	Sg 106	Bh 107	Hs 108	Mt 109									

Key:
H
hydrogen
1
← symbol
← name
← atomic number

H
hydrogen
1

Figure 5.14 The Periodic Table (elements 58–71 and 90–103 have been omitted)

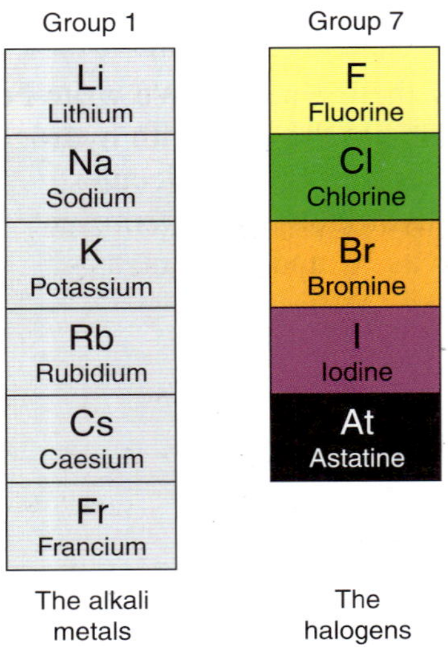

Group 1	Group 7
Li Lithium	**F** Fluorine
Na Sodium	**Cl** Chlorine
K Potassium	**Br** Bromine
Rb Rubidium	**I** Iodine
Cs Caesium	**At** Astatine
Fr Francium	
The alkali metals	The halogens

Figure 5.15 Group 1 and Group 7 in the Periodic Table

The Periodic Table is also set out so that elements with similar properties are in the same vertical column.

- The vertical columns are called **groups**. Group 1 contains the metals lithium, sodium and potassium, which have very similar properties. Group 7 contains the non-metals chlorine and bromine, which are also very similar (Figure 5.15).
- Some of the groups have names as well as numbers. These names are shown below the group numbers at the top of some columns in Figure 5.14.
- The horizontal rows in the Periodic Table are called **periods**. Period 1 contains just two elements – hydrogen and helium. Period 2 has eight elements from lithium (atomic number 3) to neon (atomic number 10).
- Non-metals are clearly separated from metals. The 20 or so non-metals are in the top right-hand corner, above the thick stepped line in Figure 5.14.
- **In each group of the Periodic Table the elements have similar properties** although there is a gradual change in properties from the top to the bottom of a group.

⓬ Look at Figure 5.15.
   a) Which element is more reactive; sodium or potassium?
   b) Which element in Group 1 do you think is i) the most reactive, ii) the least reactive?
   c) Do the elements in Group 1 get more or less reactive as you go down the group from Li to Fr?

⓭ Draw a large outline of the Periodic Table similar to Figure 5.14. On your outline, show by labelling or shading the position of:
   a) metals;
   b) the element with atomic number 14;
   c) sodium;
   d) the noble gases;
   e) the transition metals;
   f) the most reactive metal;
   g) an element used to make expensive jewellery;
   h) an element used to disinfect water supplies.

   Remember to say what the areas you have shaded represent.

5.4

Using symbols and formulae to write balanced equations

Figure 5.16 shows sparks from a sparkler. The sparks are tiny bits of burning magnesium. We are going to use symbols and formulae to write an **equation** for this reaction.

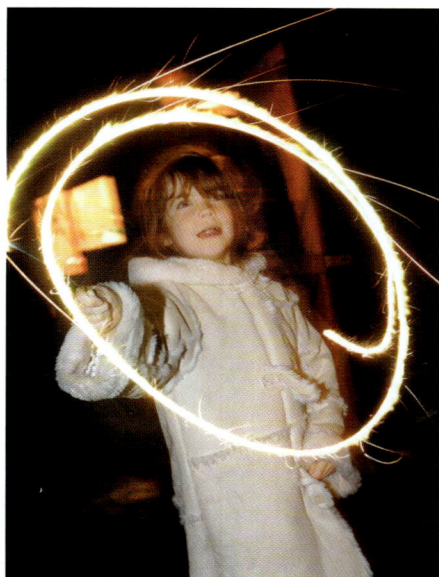

Figure 5.16 The sparks from a sparkler are tiny bits of burning magnesium.

When magnesium burns, it reacts with oxygen to form magnesium oxide. The **reactants** are magnesium and oxygen. The **product** is magnesium oxide. Here is the word equation for this reaction:

magnesium + oxygen → magnesium oxide

Chemists write symbols and formulae in equations rather than names. So instead of the word equation, we can write Mg for magnesium, O_2 for oxygen and MgO for magnesium oxide.

$$Mg + O_2 \rightarrow MgO$$

This is more helpful than the word equation, but it doesn't balance. There are two oxygen atoms in O_2 on the left, but only one oxygen atom in MgO on the right. So MgO must be doubled to give:

$$Mg + O_2 \rightarrow 2MgO$$

This equation still doesn't balance. We now have one Mg atom on the left, but two Mg atoms in 2MgO on the right. This is easily corrected by writing 2Mg on the left to give:

$$2Mg + O_2 \rightarrow 2MgO$$

The numbers of different atoms are now the same on both sides of the arrow. This is a **balanced chemical equation**.

Figure 5.17 shows a picture equation for this reaction. Picture equations help us to understand how atoms are rearranged in reactions.

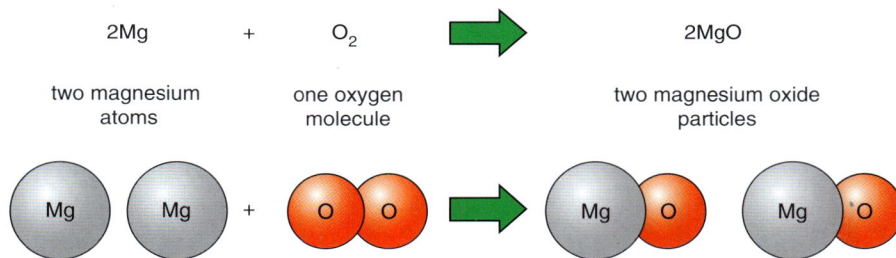

Figure 5.17 A picture equation for the reaction between magnesium and oxygen

Picture equations show that no atoms are lost during a reaction. The atoms in the reactants are still there at the end of the reaction. But the atoms are arranged differently to make new substances (the products). Because all the atoms are present at the start and end of the reaction, the mass of the products must equal the mass of the reactants.

This is summarised in the **Law of Conservation of Mass**. This says:

in any chemical change, the total mass of the products equals the total mass of the reactants.

No atoms are lost or made during chemical reactions so we can write balanced equations. In a balanced equation, there are the same number of atoms of each element on both sides of the arrow.

The following example shows the steps to follow in writing balanced equations.

Step 1 Write a word equation.

$$\text{hydrogen} + \text{oxygen} \rightarrow \text{water}$$

Step 2 Write symbols or formulae for the reactants and products.

$$H_2 + O_2 \rightarrow H_2O$$

Remember that hydrogen, oxygen, nitrogen and chlorine exist as molecules and are written as H_2, O_2, N_2 and Cl_2. All other elements are shown as single atoms (C for carbon, Fe for iron).

Step 3 Balance the equation by making the number of atoms of each element the same on both sides.

$$2H_2 + O_2 \rightarrow 2H_2O$$

Figure 5.18 shows a picture equation for this reaction.

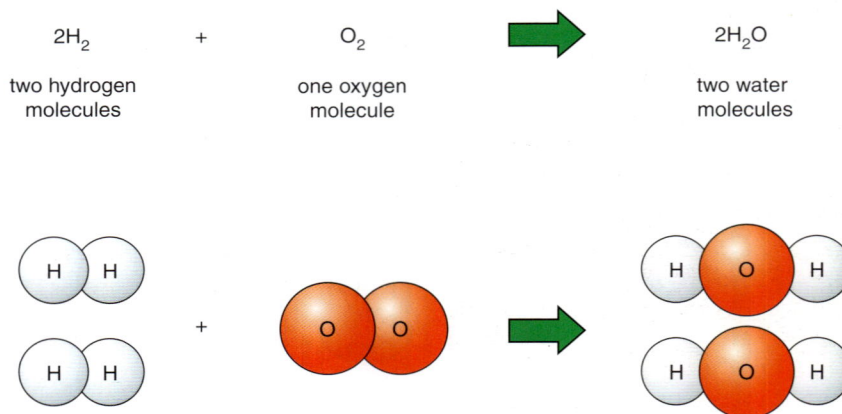

| $2H_2$ | + | O_2 | | $2H_2O$ |
| two hydrogen molecules | | one oxygen molecule | | two water molecules |

Figure 5.18 A picture equation for the reaction of hydrogen with oxygen to form water

Remember that *you must never change a formula* to make an equation balance. The formula for water is always H_2O, never HO or HO_2. The formula of magnesium oxide is always MgO, never MgO_2 or Mg_2O.

You can only balance an equation by putting numbers in front of symbols or formulae, for example 2Mg, 2MgO, $2H_2$ and $2H_2O$.

Figure 5.19 In a barbecue, charcoal (carbon) burns in oxygen to form carbon dioxide.

⓴ Look at Figure 5.19.
  a) Write a word equation for the reaction of charcoal (carbon) burning in oxygen.
  b) Copy and complete the following sentences.
     Equations must have the _____ number of atoms of each _____ on _____ sides so that the equations balance. No atoms are _____ or made during the _____.

Balanced equations are more useful than word equations because they show:
- the symbols and formulae of the reactants and products;
- the relative numbers of atoms and molecules of the reactants and products;
- the rearrangement of atoms from reactants to products.

State symbols

State symbols (s, l, g and aq) are used in equations to show the state of a substance. (s) after a formula or symbol means that the substance is a solid. (l) is used for a liquid, (g) for a gas and (aq) for an aqueous solution (a substance dissolved in water). For example,

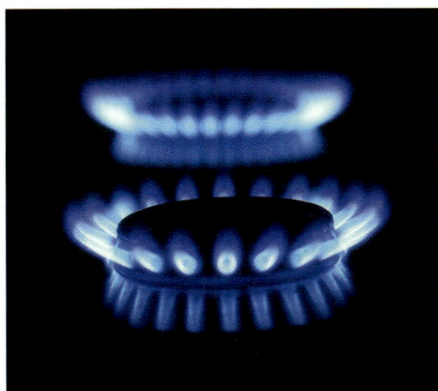

$$2Mg(s) + O_2(g) \rightarrow 2MgO(s)$$

Figure 5.20 When natural gas burns on a hob, methane (CH_4) reacts with oxygen in the air to form carbon dioxide and water.

⓯ Copy and complete the equation for the reaction in Figure 5.20. Make sure it is balanced.

$$CH_4 + \underline{\quad} O_2 \rightarrow CO_2 + \underline{\quad}H_2O$$

⓰ Propane, C_3H_8, is sold in large red cylinders. It is used as a fuel in caravans and in some homes where there is no gas pipeline. The word equation and balanced chemical equation for burning propane are shown below.

$$propane + oxygen \rightarrow carbon\ dioxide + water$$

$$C_3H_8(g) + 5O_2(g) \rightarrow 3CO_2(g) + 4H_2O(g)$$

Write down four things that the balanced equation shows but the word equation doesn't show. This will help you to see why chemists like to write balanced equations.

⓱ Copy and complete these equations. Make sure they are balanced.
  a) $\underline{\quad}Zn + O_2 \rightarrow \underline{\quad}ZnO$
  b) $N_2 + \underline{\quad}H_2 \rightarrow \underline{\quad}NH_3$
  c) $\underline{\quad}Na + \underline{\quad}H_2O \rightarrow \underline{\quad}NaOH + H_2$

5.5 # How do rocks provide building materials?

Rocks can be quarried or mined to provide essential building materials such as stone for building. One of the most important naturally occurring rocks is limestone. Limestone is mainly calcium carbonate, $CaCO_3$. This contains calcium ions, Ca^{2+}, combined with carbonate ions, CO_3^{2-}. Each carbonate ion has one carbon atom bonded to three oxygen atoms with an overall charge of $2-$.

Limestone is quarried and used as building stone. The stone can be broken into smaller pieces (aggregate and chippings) and used in concrete. Large amounts of limestone aggregate are also used for making roads.

Figure 5.21 Blocks of limestone have been used to build castles, cathedrals and houses for hundreds of years.

Figure 5.22 Mining engineers like those in these photos must survey quarry sites with great care before setting explosives. It is important to blast the rock and quarry it safely.

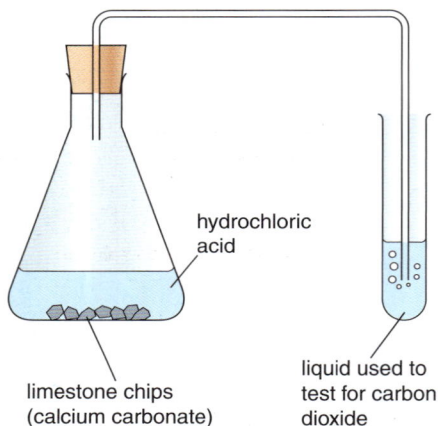

Using limestone to neutralise acidity

Limestone chippings are crushed to produce powdered limestone. This is used to neutralise acidity in soils and lakes. In this reaction, limestone (calcium carbonate) reacts with acid to produce a calcium compound, water and carbon dioxide. The formation of carbon dioxide causes the mixture to 'fizz'. For example, with hydrochloric acid:

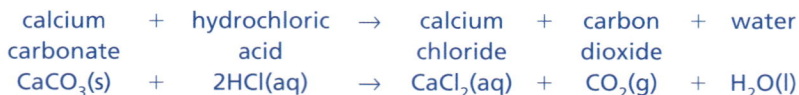

calcium carbonate	+	hydrochloric acid	→	calcium chloride	+	carbon dioxide	+	water
$CaCO_3(s)$	+	$2HCl(aq)$	→	$CaCl_2(aq)$	+	$CO_2(g)$	+	$H_2O(l)$

Figure 5.23 Calcium carbonate reacting with hydrochloric acid, followed by a test for the carbon dioxide produced

hydrochloric acid

liquid used to test for carbon dioxide

limestone chips (calcium carbonate)

⓲ a) Name the liquid used to test for carbon dioxide.
b) What happens to this liquid as carbon dioxide bubbles into it?

Decomposing limestone to make quicklime and slaked lime

Limestone (calcium carbonate) decomposes when it is heated strongly. The products are calcium oxide (commonly called quicklime) and carbon dioxide.

calcium carbonate	→	calcium oxide	+	carbon dioxide
$CaCO_3(s)$	→	$CaO(s)$	+	$CO_2(g)$

This is an example of **thermal decomposition** – using heat to break down a compound.

Carbonates of other metals decompose on heating, in a similar way to calcium carbonate. The lower a metal is in the reactivity series, the more easily its carbonate decomposes.

Calcium oxide (quicklime) reacts vigorously with water to form calcium hydroxide. Calcium hydroxide is often called slaked lime. The reaction of quicklime with water is very **exothermic** – it produces heat.

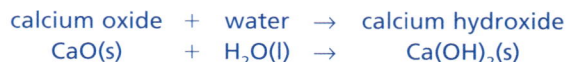

$$\text{calcium oxide} \; + \; \text{water} \; \rightarrow \; \text{calcium hydroxide}$$
$$CaO(s) \; + \; H_2O(l) \; \rightarrow \; Ca(OH)_2(s)$$

Quicklime and slaked lime are useful substances. They are used in industry as cheap alkalis to neutralise acidity. Calcium hydroxide (slaked lime) is used to make bleaching powder. Water companies use it to neutralise acid in water supplies. Farmers and gardeners also use calcium hydroxide to neutralise acidity in the soil.

Figure 5.24 summarises the reactions of calcium carbonate, calcium oxide and calcium hydroxide.

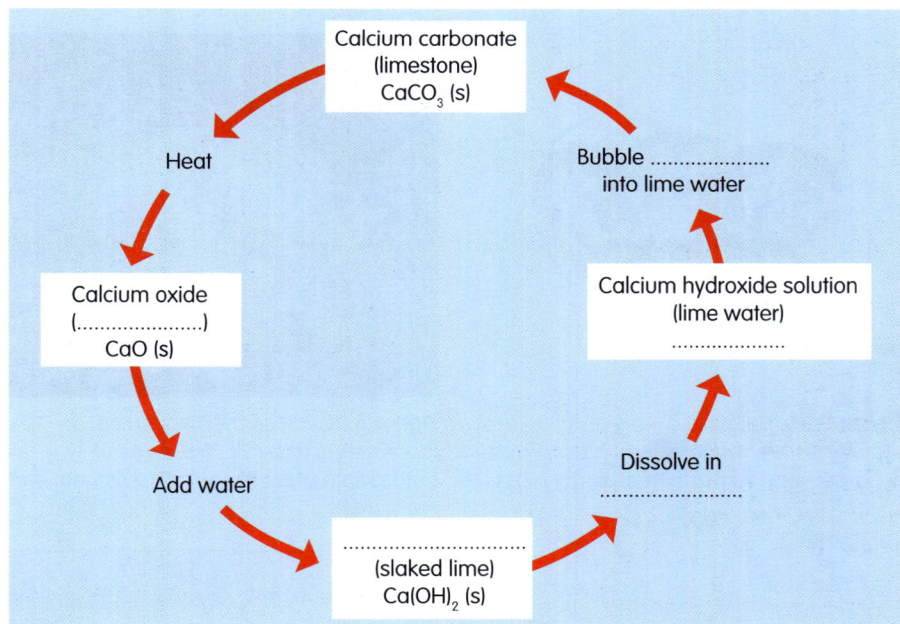

Do you remember?

Calcium hydroxide is slightly soluble in water. The solution is called lime water. This can be used to **test for carbon dioxide**. When carbon dioxide is bubbled into lime water, a milky white solid (precipitate) forms. This substance is calcium carbonate.

19 a) i) Copy and complete this equation. Make sure that it is balanced.

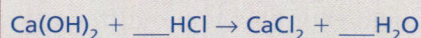

$$Ca(OH)_2 + \underline{\quad}HCl \rightarrow CaCl_2 + \underline{\quad}H_2O$$

ii) Write names for the substances in the equation.

b) Write a word equation for the action of heat on copper carbonate ($CuCO_3$).

20 Copy Figure 5.24 and fill in the blank spaces.

Figure 5.24 The reactions of calcium carbonate, calcium oxide and calcium hydroxide

How is limestone used to manufacture other useful materials?

Quicklime and slaked lime are two important products from limestone. Limestone is also used to manufacture cement, concrete and glass. It is also used in the extraction of iron from iron ore (see Section 5.7). Limestone is an extremely valuable resource for industry (Figure 5.25).

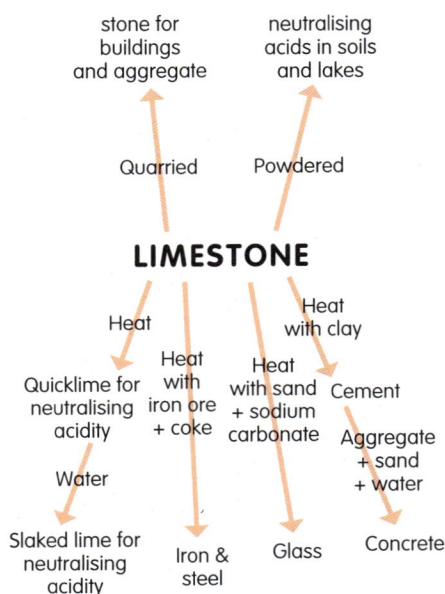

Figure 5.25 Important uses and products of limestone

Cement is made by heating limestone with clay in a kiln. When cement is used it is normally mixed with two or three times as much sand and some water. This mixture is called **mortar**. Mortar reacts slowly and sets to form a very hard material. Bricklayers use mortar to hold bricks firmly together.

Concrete is made by mixing mortar with aggregate (small pieces of broken rock). As the mortar sets around the aggregate, it produces a hard, stone-like building material. The mixture even sets under water. Isn't that strange?

Glass is made by heating a mixture of metal oxides or metal carbonates with pure sand (silicon dioxide, SiO_2) in a furnace. At the high temperatures in the furnace, carbonates decompose to oxides, and bubbles of carbon dioxide form in the mixture. If the mixture is heated further, the bubbles escape and a runny liquid forms. As this liquid cools, it becomes thick enough to be moulded or blown into different shapes. On further cooling, the glass sets solid.

Ordinary glass for bottles and windows is made by heating a mixture of limestone (calcium carbonate), soda (sodium carbonate) and sand. This is sometimes called soda glass.

Figure 5.26 Cutglass dishes and ornaments are made of lead glass, which is harder and shinier than ordinary glass.

Figure 5.27 Glass ovenware and laboratory glassware are made of borosilicate glass (Pyrex®). This is heat resistant.

Figure 5.28 Blue glass is made by adding cobalt oxide to the usual reactants for making glass.

21 Three types of special glass are shown in Figures 5.26, 5.27 and 5.28.
   a) What substance do you think is added to the usual reactants to make lead glass?
   b) Which element produces the colour in blue glass?
   c) Ordinary glass has been improved and developed to produce other types of glass. Write down three improvements or developments that these photos show.
   d) Write down two risks in the use of glass.

Activity – Quarrying or countryside?

Read the following passage and then answer the questions.

The quarrying and mining of rocks creates some benefits but also has some drawbacks. These are well illustrated by the quarrying of limestone in the UK. In Britain, limestone is found in some of the most beautiful areas – the Yorkshire Dales, the Peak District in Derbyshire, the Chilterns in Buckinghamshire and the Sussex Downs. The quarrying of limestone can spoil the countryside and destroy wildlife habitats.

On the other hand, limestone is a very important raw material for industry. Every year about 90 million tonnes of limestone are quarried in Britain. The limestone industry provides useful products. It also provides jobs for people and increases the country's wealth.

These benefits of quarrying must be balanced against the problems it creates.

Look carefully at Figure 5.29. It shows the quarry near Littleham.

❶ Write down two lists. The first list should show three benefits (advantages) that the quarry brings to Littleham. The second list should show three drawbacks (disadvantages).

❷ Suppose you live in Littleham. The quarry operators are called Limestone UK. Write a letter to the Chief Executive of Limestone UK. Tell him about the complaints that the people who live in Littleham have about the quarry.

❸ Suppose you are the Chief Executive of Limestone UK. Write a reply to the people who live in Littleham. Tell them about your efforts to reduce problems from the quarry. Mention benefits that the quarry has created for the people of Littleham.

❹ Science can help us in many ways. For example, it can tell us how to make glass from limestone. But there are some questions that science cannot answer. This may be because people's beliefs or personal views are important. Or it may be because we cannot find reliable evidence. Science cannot tell us whether the quarry at Littleham has given the people who live there a better lifestyle. Why not?

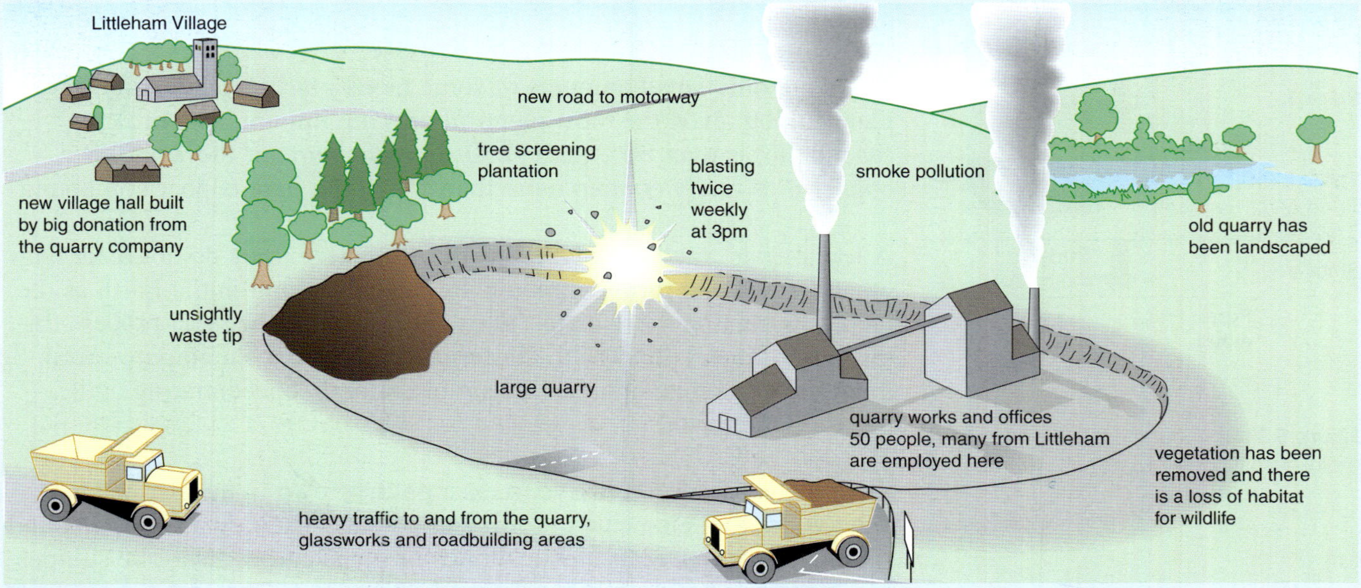

Littleham Village

new road to motorway

tree screening plantation

blasting twice weekly at 3pm

smoke pollution

old quarry has been landscaped

new village hall built by big donation from the quarry company

unsightly waste tip

large quarry

quarry works and offices 50 people, many from Littleham are employed here

vegetation has been removed and there is a loss of habitat for wildlife

heavy traffic to and from the quarry, glassworks and roadbuilding areas

Figure 5.29 The quarry and surrounding area near Littleham

5.6 How do rocks provide metals?

Rocks provide important **metal ores** as well as building materials. These metal ores contain enough metal to make it worthwhile to extract the metal (Table 5.2).

Name of ore	Name and formula of metal compound in the ore	Metal obtained
bauxite	aluminium oxide, Al_2O_3	aluminium
iron ore (haematite)	red iron oxide, Fe_2O_3	iron
copper pyrites	copper sulphide, CuS and iron sulphide, FeS	copper

Table 5.2 The ores from which we get aluminium, iron and copper

Figure 5.30 Attractive crystals of gold on quartz

From ores to metals

Extracting (getting) metals from their ores usually involves four stages (Figure 5.31):

1 mining (digging up) the ore;
2 concentrating (separating) the ore;
3 reducing (converting) the ore to the metal;
4 purifying the metal.

Mining metal ores involves quarrying, tunnelling or open-cast digging. After mining, the ore is separated from impurities such as soil and waste rock. This is called **concentrating** the ore. First the rock is crushed. Then the ore is separated from the waste by using their different densities.

There are only small amounts of some metal ores. Others are more plentiful, but even the most plentiful ores are impure. Iron ore (haematite) is over 80% pure Fe_2O_3 in many parts of the world. But copper ores rarely contain more than 1% of the pure copper compound.

A few metals, such as gold, are very **unreactive** (they do not react with other elements to form compounds). They are found in the Earth as the metal. Extracting these metals does not involve a chemical process. In the case of metals like gold, the concentrated ore is the impure metal. So these metals can be obtained by simply mining, concentrating and purifying the metal.

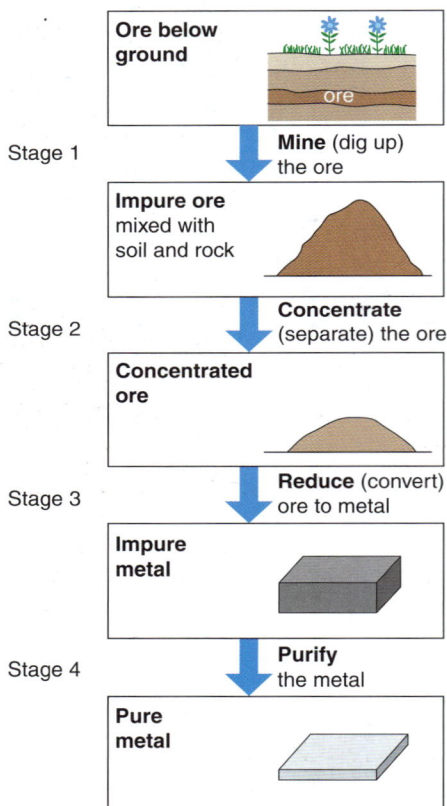

Figure 5.31 The four stages in extracting a metal from its ore

Most metals are too **reactive** to exist on their own in the Earth. They react with other elements to form compounds. Their ores are compounds – usually metal oxides, metal sulfides or metal carbonates. The sulfides and carbonates can easily be changed into oxides. The metal can then be obtained from the metal oxide by removing oxygen. When the metal oxide loses its oxygen, the process is called **reduction**. We will study reduction in more detail in Section 5.7.

Sodium — Most reactive
Calcium
Magnesium
Aluminium
Carbon
Zinc
Iron
Lead
Copper
Silver
Gold — Least reactive

Decreasing reactivity

Figure 5.32 The reactivity series, showing the position of carbon

There are two important methods of reducing metal compounds. These methods depend on how reactive the metal is. The **reactivity series** (Figure 5.32) shows the metals in order of their reactivity.

Carbon is *not* a metal, but it has been included in Figure 5.32. Any element higher in the reactivity series can be used to **displace** (separate) a lower element from its compounds.

Metals in the middle of the reactivity series, such as zinc and iron, are below carbon (see Figure 5.32). They are less reactive than carbon. This means that carbon can be used to displace (separate) them from their compounds. So zinc and iron can be extracted by reducing their oxides with carbon (coke) or carbon monoxide.

Figure 5.33 The ruins of a tin mine in Cornwall. Tinstone (tin oxide, SnO_2) was once mined in Cornwall. It was reduced to tin using carbon (coke) in furnaces near the mines. We only mine metal ores if selling the metal makes more money than it costs to extract the metal. This mine closed because it did not make a profit. If the price of tin increased greatly, tin mining might start again in Cornwall.

22 a) Copy and complete the following word equation for the formation of tin.

   tin oxide + carbon →

   b) Why is tin no longer produced in Cornwall?

23 a) Name two metals (apart from zinc and iron) that can be obtained by reducing their oxides with carbon (coke) or carbon monoxide.
   b) Name two metals (apart from sodium and aluminium) that can be obtained by electrolysis of their molten compounds.

Metals near the top of the reactivity series, such as sodium and aluminium, often occur as oxides or chlorides. These metals are above carbon in the reactivity series. So carbon cannot be used to extract these metals from their oxides. Instead, these metals are extracted by **electrolysis**. This involves decomposing the molten oxide or chloride to the metal using electricity.

Counting the cost of extracting metals

Extracting metals from their ores involves turning huge quantities of raw materials into useful and valuable metals. These metals are then used to manufacture a vast range of products – vehicles, tools, pans, cutlery, cans, jewellery, pipes and girders.

jobs with good salaries

benefits

metal ore

drawbacks

pollution and noise

Figure 5.34 One of the benefits and one of the drawbacks of mining ores and extracting metals

24 Look at Figure 5.34.
a) Write down two other benefits of mining ores and extracting metals.
b) Write down two other drawbacks of mining ores and extracting metals.
c) Sketch Figure 5.34 and add your own benefits and drawbacks to it.

The extraction of metals followed by the production and sale of valuable metal products creates jobs for many people. This improves their standard of living and adds to the wealth of a nation. But we don't get these benefits for nothing. There are costs – social, environmental and economic.

Social costs

There are health and safety risks for the people who work in mining and metal industries. Some processes may involve poisonous chemicals. Loud factory noise can harm a worker's hearing. People living in the area near mines and factories can also be affected.

But if people follow health and safety rules, they can work safely, even with dangerous chemicals and in noisy factories.

Large industrial operations like mining and quarrying often bring many more people to live and work in the area. This puts a strain on the schools and hospitals because, at one time, there were fewer people.

Environmental costs

Wildlife habitats and farmland are often destroyed in areas where there are mines and quarries. The mining, quarrying and transport of materials also creates noise and pollution. There are various forms of pollution. Factory chimneys and vehicle exhausts produce air pollution. Dust from blasting also pollutes the air. Unsightly tips of waste material spoil the environment.

Economic costs

Large industrial operations are also very costly. They need expensive machinery and fuels. Fuels are needed to operate machines, heat buildings and maintain chemical processes. They are also needed to transport workers and materials.

When you next buy a can of Coke® or a piece of jewellery, remember the costs. These items may add to your enjoyment, but their production has social, environmental and economic costs.

5.7 How is iron extracted from iron ore?

The main raw material for making iron is iron ore (haematite). Haematite is impure iron oxide, Fe_2O_3. The iron ore is usually obtained by open-cast mining.

Iron is extracted from the iron ore in a **blast furnace**. Blast furnaces are built as towers about 15 metres tall. Figure 5.35 on the next page shows a diagram of a blast furnace with an explanation of the processes at the side.

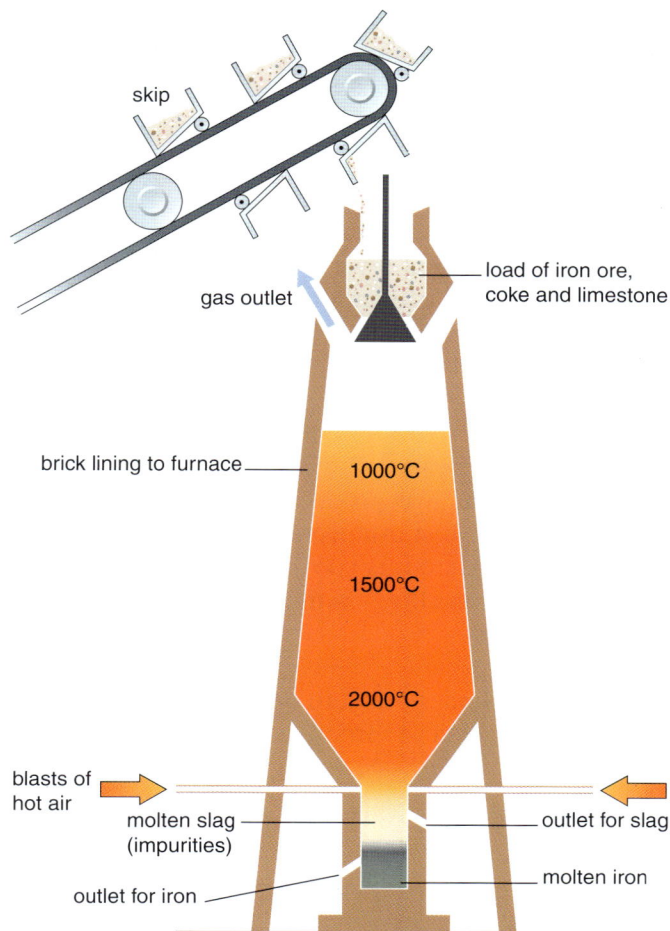

Figure 5.35 Extracting iron from iron ore in a blast furnace

1 Solid raw materials (iron ore, coke (carbon) and limestone) are added at the top of the furnace.

2 Blasts of hot air (which give the furnace its name) are blown in near the bottom of the furnace.

3 Oxygen in the blasts of air causes the coke (carbon) to burn, forming carbon dioxide.

$$\begin{array}{ccccc} \text{carbon} & + & \text{oxygen} & \rightarrow & \text{carbon dioxide} \\ \text{C} & + & \text{O}_2 & \rightarrow & \text{CO}_2 \end{array}$$

4 At the high temperatures in the furnace, carbon dioxide reacts with more coke (carbon) to form carbon monoxide.

$$\begin{array}{ccccc} \text{carbon dioxide} & + & \text{carbon} & \rightarrow & \text{carbon monoxide} \\ \text{CO}_2 & + & \text{C} & \rightarrow & \text{2CO} \end{array}$$

5 The carbon monoxide reacts with the iron ore (red iron oxide) producing carbon dioxide and molten iron.

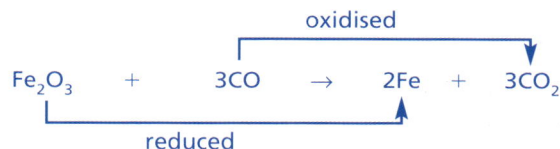

red iron oxide + carbon monoxide → iron + carbon dioxide

$$\underset{\text{reduced}}{\underbrace{\text{Fe}_2\text{O}_3}} + \overset{\text{oxidised}}{3\text{CO}} \rightarrow 2\text{Fe} + 3\text{CO}_2$$

6 Molten iron runs to the bottom of the furnace and is tapped off from time to time.

In the reaction between Fe_2O_3 and CO, carbon monoxide gains oxygen, forming carbon dioxide. This gain of oxygen is called **oxidation.** The carbon monoxide is said to be **oxidised.** At the same time, iron oxide loses oxygen. This loss of oxygen is called **reduction**. The iron oxide is said to be **reduced.**

Oxidation and reduction always occur together. If one substance gains oxygen and is oxidised, another substance must lose oxygen and be reduced. We call the combined process **redox** (**red**uction + **ox**idation).

25 The reaction between carbon dioxide and coke (carbon) to form carbon monoxide is a redox reaction.
   a) What is a redox reaction?
   b) In the above redox reaction, which substance is:
      i) oxidised, ii) reduced?

Why is limestone used in the furnace?

The main impurity in iron ore is sand (impure silicon dioxide, SiO_2). This is removed by limestone.

Figure 5.36 Molten iron being poured from a furnace

At the high temperatures in the furnace, limestone decomposes forming calcium oxide and carbon dioxide. The calcium oxide, which is a base, reacts with sand (SiO_2), which is acidic, to form 'slag', calcium silicate.

$$\text{calcium oxide} + \text{silicon dioxide} \rightarrow \text{calcium silicate}$$
$$CaO + SiO_2 \rightarrow CaSiO_3$$

Molten 'slag' falls to the bottom of the furnace and floats on the molten iron. This can be tapped off at a different level from the molten iron. The 'slag' is used in road making and cement manufacture.

Why is iron converted to steel?

Iron from the blast furnace contains about 96% iron. The main impurity in this iron is carbon. This makes it brittle, so it is not very useful. Removing all the impurities from the iron produces pure iron. However, pure iron is too soft for most uses. Most iron from the blast furnace goes straight to a steel-making furnace. Here it is converted into steel, which has strength and hardness. Steel is an alloy – a mixture of iron with carbon and sometimes other metals.

Steel is made by blowing oxygen under pressure onto the hot, molten impure iron. The oxygen converts excess carbon to carbon dioxide which escapes as a gas.

Using alloys

Alloys are made by melting the main metal and then dissolving other elements in it.

Alloys can be made to have particular properties for special uses. Some are particularly hard. Some are resistant to corrosion. Others have special magnetic or electrical properties.

The most important alloys are steels. The composition, properties and uses of various steels are shown in Table 5.3.

Type of steel	Composition	Properties	Uses
Low-carbon steel (mild steel)	99.8% iron 0.2% carbon	Easily pressed into shapes	Car bodies
High-carbon steel	98.0% iron 1.7% carbon 0.3% manganese	Hard but brittle	Tools
Stainless steel	73.7% iron 0.3% carbon 18.0% chromium 8.0% nickel	Hard and resistant to corrosion	Cutlery, pans

Table 5.3 The composition, properties and uses of various steels

Most metals in everyday use are alloys. Pure copper, pure aluminium and pure gold are too soft for most uses. So they are mixed with small amounts of other metals to make them harder. During the last 40 years, aluminium alloys have been used more and more. These include duralumin, which contains 4% copper. Aluminium alloys are light, strong and corrosion resistant. They are used for aircraft bodywork, overhead electricity cables and lightweight tubing.

In recent years, **smart alloys**, sometimes called 'memory metals', have been developed. These can return to their original shape after being deformed. Smart alloys are used in spectacle frames and in dental braces. Smart alloy dental braces are made to return to their original shape after fitting. This pulls the teeth into improved positions in the gum.

Many important alloys have **transition metals** as their main elements. Transition metals occupy the central block of the Periodic Table (Figure 5.38). They include iron, copper, chromium, nickel and titanium.

Transition metals, like all metals, are good conductors of heat and electricity. They can also support heavy loads and can be bent or hammered into shape. So transition metals are useful for building metal structures. They are also useful for things that must conduct heat or electricity easily. Steel is one of our most important structural materials. It is used in girders, joists and bridges. Copper has properties that make it useful for electrical wiring and in pipes for plumbing.

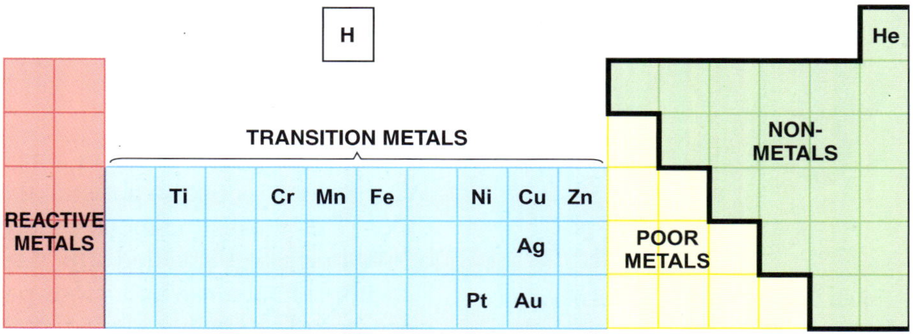

Figure 5.37 Spectacle frames made of 'smart alloys' return quickly to their original shape after being bent.

Figure 5.38 The position of the transition metals in the Periodic Table

Figure 5.39 A hip joint made of titanium alloy

Aluminium and titanium are two other useful metals because of their low density and resistance to corrosion. Both metals occur as their oxides in ores. But these oxides cannot be extracted by reduction with carbon. Present-day methods of extracting the two metals are expensive. This is because:
- large amounts of energy are needed;
- there are several stages in the process.

These costs limit the use of titanium.

26 This question is about new ways to extract copper. Answer the questions as you read the different parts of the passage.

Copper ores contain copper sulfide (CuS). At one time, all copper was extracted from these sulfide ores by first converting the copper sulfide to copper oxide. This copper oxide was then reduced to copper by heating with carbon.

During the last 20 years, good quality (copper-rich) ores have become scarce. This has led chemists to look for new ways of extracting copper from poor-quality (low-grade) ores.
a) What is the major problem of mining low-grade ores which contain massive amounts of useless rock?
Fortunately, there are helpful bacteria in most copper ores. These bacteria can oxidise insoluble copper sulfide to soluble copper sulfate (CuSO₄) using oxygen in the air.

b) Write a word equation for this reaction.

The bacteria in copper ores can survive in harsh conditions:
- the high acidity caused by dilute sulfuric acid, which is used in the process;
- the higher temperatures caused by the reaction;
- the presence of poisonous copper compounds, which would kill most other organisms.

A heap of crushed rock is sprayed with very dilute sulfuric acid. Dilute copper sulfate solution trickles from the bottom of the pile. Finally, copper is extracted from the copper sulfate solution by electrolysis.
c) Which of the three conditions mentioned above could help the chemical process to go faster?
d) Which of the three conditions mentioned above could damage plants and animals?

Explaining the properties of alloys

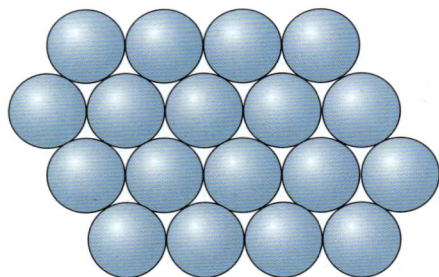

Figure 5.40 Close packing of atoms in a metal

Using X-rays, scientists have shown that the atoms in most metals are packed as close together as possible. This arrangement is called **close packing**. Figure 5.40 shows a few close-packed atoms in one layer of a metal.

The bonds between atoms in a metal are strong, but they are not rigid. When a force is put on a metal, the layers of atoms can slide over each other. This means the metal is soft and malleable. This movement of atoms in a metal is called **slip**. After slip has happened, the atoms settle into position again with a close-packed structure. Figure 5.41 shows the positions of atoms before and after slip. This is what happens when a metal is hammered or pressed into different shapes.

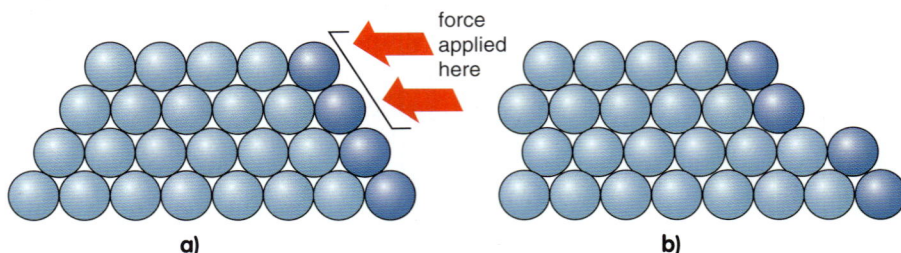

Figure 5.41 The positions of atoms in a metal a) before and b) after 'slip' has occurred

force applied here

In steel, the close-packed arrangement of iron atoms is broken up by smaller carbon atoms. The different-sized carbon atoms break up the layers of iron. This makes it more difficult for the layers to slide over each other. So steel alloy is harder and much less malleable than pure iron (Figure 5.42).

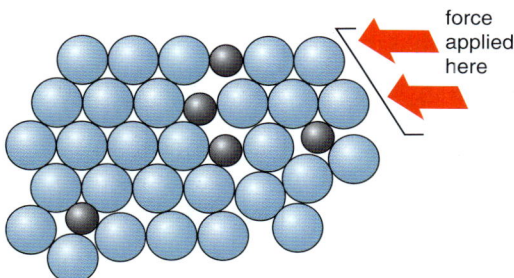

Figure 5.42 Slip cannot occur so easily in steel alloys because the atoms of different sizes cannot slide over each other.

Activity – Recycling metals

Read the passage below and then answer the questions.

Scrap metals such as empty drinks cans, used aluminium foil and broken machines can be melted down and used again. Local authorities are now expected to achieve certain targets for recycling waste materials such as metals, glass, paper and plastics. In many areas, there are strong reminders to help us recycle these materials.

Recycling metals is not easy. Scrap metal has to be collected. Then it has to be transported to where it can be re-used. The metals have to be separated from other materials and sometimes one metal has to be separated from another.

1. Which of the four stages in extracting metals (Figure 5.31) are avoided if metals are recycled?
2. Write down a list of three savings or advantages of recycling metals.
3. a) How do you think metals are separated from paper? Bear in mind that the metal has to be melted down during its re-use.
   b) The two metals recycled in the largest amounts are aluminium and iron (steel). How do you think these are separated?
4. Why is almost all the gold we use recycled, but only about half the aluminium?
5. Find out about the plans and targets for recycling metals in your area. Write a few sentences about them.
6. What further action could be taken to improve the recycling of metals in your area?

You can read more about recycling in Chapter 3.

㉗ a) Why are the furnaces used to make iron called *blast* furnaces?
   b) Why is most iron from the blast furnace converted to steel?
   c) Write down three reasons why steel (iron) is used in greater quantities than any other metal.

㉘ The strengths of two metal wires can be compared by measuring the force needed to break each wire. The apparatus which might be used is shown in Figure 5.43.
   a) Describe briefly how you would carry out the experiment.
   b) What measurements would you make?
   c) Copy and complete the sentences below, using the words from this box.

dependent	fair	independent
measured	same	selected

In this experiment the _____ variable is the type of metal wire used. This is the variable that is changed or _____ by the investigator.

The _____ variable in the experiment is the force needed to break each wire. This is the variable that is _____ when the independent variable changes.
In order to make the experiment _____, it is important that only the independent variable affects the dependent variable. All other possible variables must be kept the _____.
   d) State one variable that you would control to ensure the wires are tested fairly.

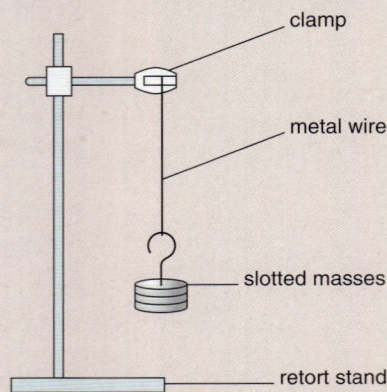

Figure 5.43 Comparing the strengths of metal wires

Summary

✓ **Chemistry** is the study of materials and substances. Chemists and chemical engineers study materials and try to change **raw materials**, such as iron ore and limestone, into useful **manufactured materials**, such as steel, cement and glass.

✓ **Elements** are the simplest substances. They cannot be broken down into simpler substances.

   Compounds are substances containing two or more elements chemically combined together.

   Mixtures are two or more substances which are mixed together but *not* combined chemically.

   An **alloy** is a mixture of a metal with one or more other elements.

✓ All substances are made of **atoms**. Each element contains only one sort of atom. Compounds contain two or more different atoms joined together by chemical bonds.

✓ When atoms combine, they can either:
   ● share electrons to form **molecules**,
   ● give and take electrons to form **ions**.
   An **atom** is the smallest particle of an element.

   A **molecule** is a particle containing two or more atoms joined by chemical bonds.

   An **ion** is a charged particle formed from an atom by the loss or gain of one or more electrons.

✓ Elements can be represented by **symbols**. Compounds can be represented by **formulae**. A formula shows the number of atoms of the different elements that are joined together in one molecule of a compound.

✓ In the Periodic Table, elements are arranged in order of their atomic number. In each group of the Periodic Table, the elements have similar properties.

✓ **Balanced equations** use symbols and formulae to summarise the reactants and products in a reaction.

✓ **Limestone** is mainly calcium carbonate, $CaCO_3$. It is one of the most useful naturally-occurring building materials. It is used to manufacture quicklime, slaked lime, cement, concrete and glass.

✓ **Extracting metals** from their ores involves four stages:
  - mining the ore;
  - concentrating the ore;
  - reducing the ore to the metal;
  - purifying the metal.

✓ There are two main methods of extracting metals from their ores:
  - electrolysis of molten compounds for metals above carbon in the reactivity series;
  - reduction of oxides with coke (carbon) or carbon monoxide for metals below carbon in the reactivity series.

✓ Quarrying, mining and metal industries bring both benefits and problems (Table 5.4).

Benefits	Problems
• Produces useful products and materials • Creates jobs and employment • Increases the wealth of a community	• Causes pollution from smoke, dust and noise • Destroys wildlife habitats • Spoils the environment with waste from quarries and mines

Table 5.4 The benefits and problems of quarrying, mining and metal industries

✓ **Redox** involves reduction and oxidation.

  Oxidation occurs when a substance gains oxygen.

  Reduction occurs when a substance loses oxygen.

✓ Recycling metals is important because:
  - it helps to save limited resources of the metal ores;
  - it reduces damage to the environment;
  - it saves the cost of extracting the metals.

EXAMQUESTIONS

❶ At one time, gutters and drainpipes were made of iron. Today they are made of plastics like PVC (polyvinyl chloride).
  a) State two properties of iron which made it useful for gutters and drainpipes. (*2 marks*)
  b) State one problem of using iron. (*1 mark*)
  c) Why has iron been replaced by plastics? (*2 marks*)

❷ a) Limestone is used to make some important products. Name two important products from limestone listed in the box below. (*2 marks*)

> alcohol   concrete   diesel   glass   PVC

  b) In an experiment, a student heated a piece of limestone very strongly, as shown in Figure 5.44.

i) State *one* safety precaution that the student should take when doing this experiment. (*1 mark*)

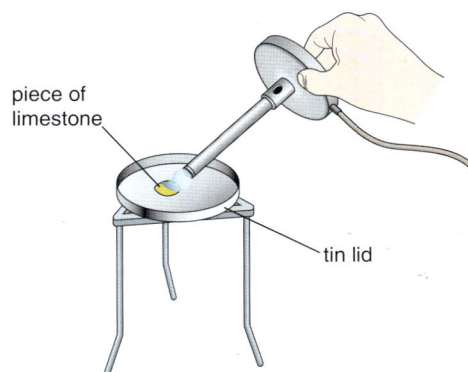

piece of limestone

tin lid

Figure 5.44

ii) When limestone is heated, it forms two products: a white powder and carbon dioxide. What is the chemical name of the white powder? (*1 mark*)

c) The student did a second experiment using 3.00 g of limestone. The limestone was weighed before and after being heated. The student then repeated this experiment with a new sample of 3.00 g of limestone. The results of the experiment are shown in the table below.

	Experiment 1	Experiment 2
Mass of limestone before heating (g)	3.00	3.00
Mass of limestone after heating (g)	1.68	1.72
Mass lost (g)	1.32	1.28

i) What is the average mass lost for the two experiments. *(1 mark)*
ii) Why is it important to repeat the experiment? *(1 mark)*
iii) Why is the mass lost not the same for the two experiments? *(1 mark)*
iv) Why is a balance which measures to the nearest 0.1 g not suitable for this experiment? *(1 mark)*

❸ Copy and complete the sentences below, using words from this box.

> atoms   close   different   forced
> metal   move

a) Close packing describes _____ structures in which the atoms are packed together as _____ as possible. *(2 marks)*
b) Slip occurs when the _____ in a metal _____ from one position to another. *(2 marks)*
c) Malleable describes the way in which a material can be _____, pressed or hammered into _____ shapes. *(2 marks)*

❹ Look at Figure 5.45.

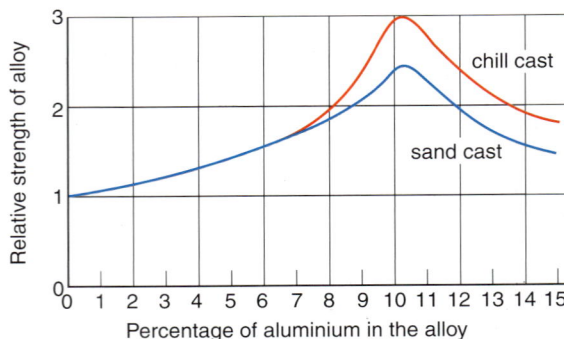

Figure 5.45 The effect of aluminium on the strength of copper alloys

a) Which process produces the strongest alloy – chill casting (rapid cooling) or sand casting (slow cooling) of the liquid alloy? *(1 mark)*
b) What percentage of aluminium produces the strongest alloy? *(1 mark)*
c) How many times stronger is this alloy than pure copper? *(1 mark)*
d) What percentage of aluminium produces a sand-cast alloy twice as strong as pure copper? *(1 mark)*
e) Are chill casting and sand casting examples of ordered variables or categoric variables? *(1 mark)*

❺ a) A word equation for making a simple glass of calcium silicate ($CaSiO_3$) is:

limestone + sand → calcium silicate + carbon dioxide

Write a balanced chemical equation for the reaction. *(2 marks)*
b) Glass is used for bottles and ovenware.
  i) List two properties which make glass very suitable for these uses. *(2 marks)*
  ii) List two drawbacks of using glass for bottles and ovenware. *(2 marks)*

❻ Quarrying has an effect on the environment and on people's lives. Write down five important points about the social, economic, health or environmental effects of quarrying. *(5 marks)*

Chapter 6
How does crude oil provide useful materials?

At the end of this chapter you should:

✓ understand that crude oil is a mixture of useful materials which are mainly alkanes;

✓ know about the processes at oil refineries;

✓ know that burning fossil fuels causes air pollution such as acid rain and climate change;

✓ know how cracking is used to make additional amounts of petrol;

✓ know that cracking produces materials called alkenes;

✓ know that alkenes are used to make plastics which are polymers;

✓ know that the polymers made from alkenes have a range of useful properties;

✓ know that some plastics can be recycled easily;

✓ understand that most plastics are not biodegradable (do not rot away);

✓ be able to judge the effect of fuels on the environment;

✓ know about the development of better fuels and smart plastics;

✓ understand that ethanol (spirit fuel) can be manufactured from renewable and non-renewable sources.

Figure 6.1 In the twenty-first century most garages don't mend cars. The garage is where you get your milk, lottery ticket, newspaper, sweets, phone top-up card – and fuel for the family car. Garages are part of our everyday lives. Crude oil provides the fuel we buy at garages and many more useful materials.

6.1 Crude oil

Crude oil is a naturally occurring material. It is found under the ground. It is a mixture of similar compounds called **hydrocarbons**. Petrol, candle wax and polythene also contain hydrocarbons. Hydrocarbons are compounds that contain only hydrogen atoms and carbon atoms. Crude oil was made from living organisms. These organisms died and rotted to make the oily materials. Crude oil is sticky stuff. It often gets onto other things when it is spilled.

Distillation

Mixtures of liquids can be separated by **distillation**.

A **hydrocarbon** is a compound that contains hydrogen and carbon atoms only.

Distillation is when a liquid is boiled and the vapour that boils off is cooled and condensed back to a liquid.

1. How do you think you could separate the following mixtures?
   a) crude oil and seawater, to get clean seawater;
   b) crude oil and seawater, to recover crude oil;
   c) crude oil and rocks, to get clean rocks. (You may need to use another chemical. What will wash off oily stuff?)

2. If you had to clean oil off a bird, what would you use?

3. How would the bird react to being cleaned in this way?

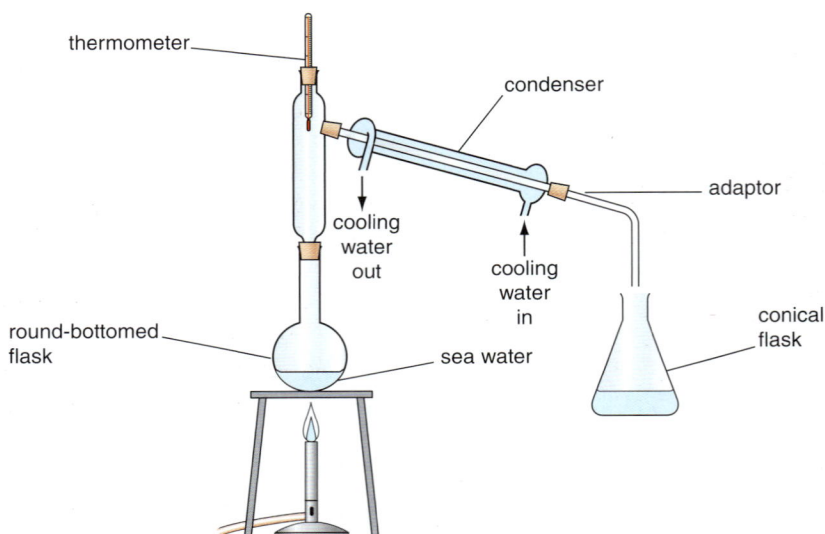

Figure 6.3 A pure liquid can be separated out of a solution using this apparatus. Water or air can be used to cool the condenser. In this case, the condenser is cooled by water.

Figure 6.2 This bird may not like this treatment, but it will die if it does not receive it.

Different substances have different boiling points. When a liquid boils, the particles become separated from each other and form a gas. The boiling point of the liquid depends on:

- atmospheric pressure – substances boil at a lower temperature when the atmospheric pressure is lower;
- the mass of the molecules in the liquid – heavier molecules need more energy to separate them.

If you boil salty water (sodium chloride solution), only the water molecules are turned into vapour. The sodium and chloride particles in the water do not vaporise. They remain in the boiling liquid. The vapour that boils off is pure water vapour. This can be cooled and condensed to produce pure water. The process is called distillation. You can use distillation to make pure water from salt water.

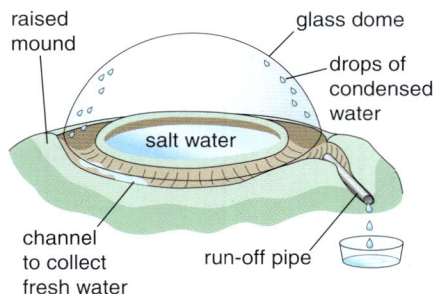

Figure 6.4 Turning salt water into pure water – sunshine is used as an energy source.

Fractional distillation

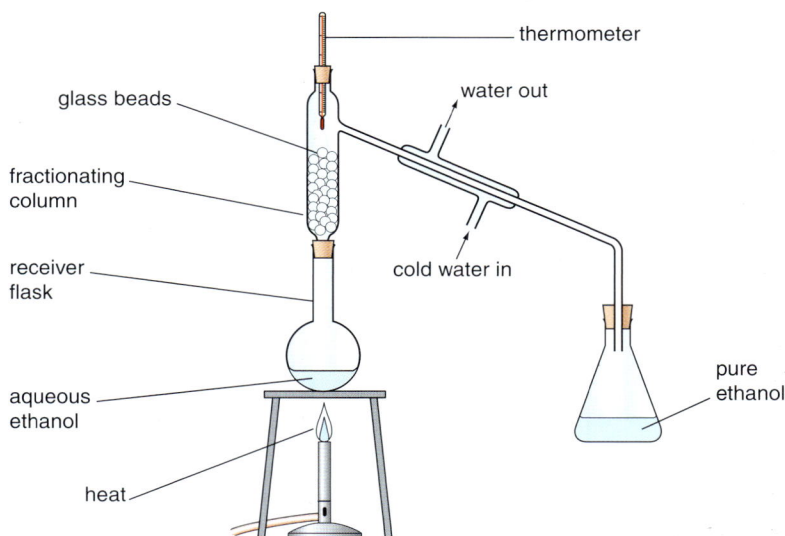

Figure 6.5 During fractional distillation, the vapours condense and evaporate several times in the fractionating column.

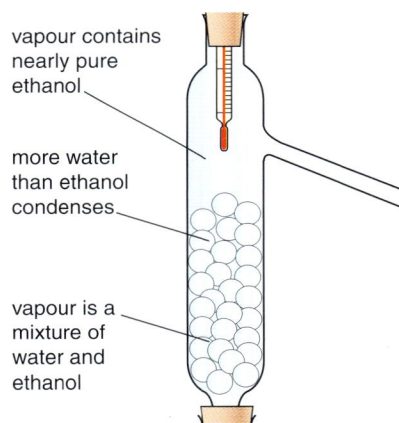

Figure 6.6 How a fractionating column works

Ordinary distillation cannot separate liquids with boiling points that are close together. This is done by **fractional distillation**.

When a mixture of ethanol (boiling point 78 °C) and water (boiling point 100 °C) is boiled, both the ethanol and the water molecules vaporise together. When the vapour is condensed, it is still a mixture. But there is more ethanol than water because ethanol boils at a lower temperature.

However, if you vaporise and condense the boiling mixture several times, then more and more ethanol is vaporised and more water remains as liquid. This is how a vertical fractionating column works.

The liquid evaporates and condenses several times on its way up the column. Then only vapour of the liquid with the lower boiling point reaches the top. This is cooled, condensed and collected.

6.2 Alkanes – the simplest hydrocarbons

Crude oil is a mixture of materials which are mainly **alkanes**. Alkanes are simple hydrocarbons. Carbon atoms can form four chemical bonds but hydrogen atoms can form only one chemical bond. So the simplest alkane of all is methane, CH_4, as shown in Figure 6.7.

A hydrocarbon with two carbon atoms is shown in Figure 6.8, page 106. It is called ethane. Ethane is also an alkane.

Ethane is like methane with a $-CH_2-$ unit added. If you add another $-CH_2-$ unit you get the next alkane called propane (Figure 6.9a).

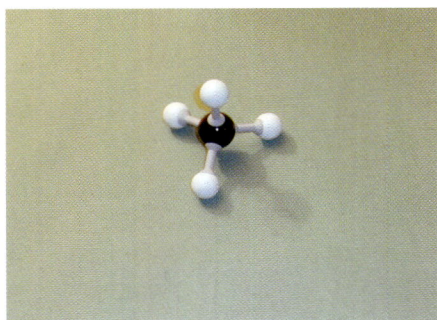

Figure 6.7 The structure of methane

Adding another –CH$_2$– unit makes butane (Figure 6.9b).

By adding a –CH$_2$– unit each time, we get a series of similar molecules. This series of hydrocarbons makes up the alkanes. All alkanes have similar properties. Their most important property is that they burn in air to form carbon dioxide and water. When they burn, they also give out lots of heat. So alkanes are used as fuels (Table 6.1).

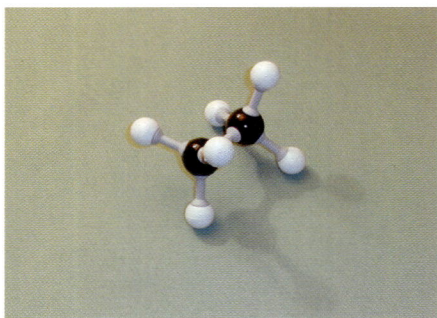

Figure 6.8 The structure of ethane

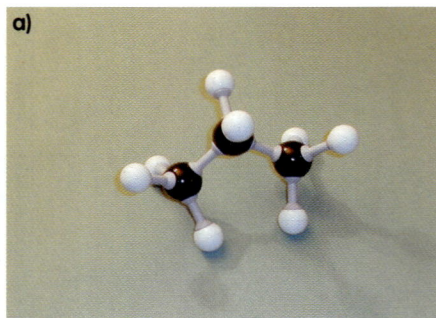

Figure 6.9 The structure of a) propane and b) butane

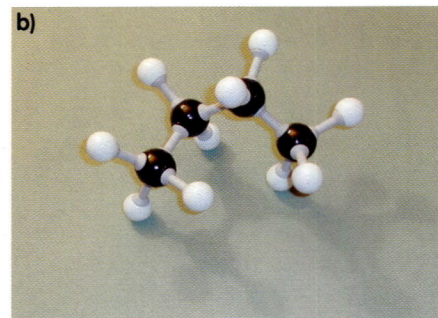

Alkane	Use
Methane	Kitchen gas for cookers
Propane	Camping gas (calor gas) for cookers
Butane	Bottled gas for heating
Octane	Petrol for cars
Dodecane	Diesel for vans

Table 6.1 The use of alkanes as fuels

Activity – Alkanes and boiling points

Is there a link between the size of the molecules in an alkane and its boiling point?

Alkane	Number of carbon atoms in a molecule	Boiling point (°C)	Alkane	Number of carbon atoms in a molecule	Boiling point (°C)
Methane	1	−161	Hexane	6	69
Ethane	2	−89	Heptane	7	98
Propane	3	−42	Octane	8	125
Butane	4	0	Nonane	9	151
Pentane	5	36	Decane	10	174

Table 6.2 Alkanes and their boiling points

❶ Plot a graph of the boiling points for these alkanes against the number of carbon atoms in the molecules.
❷ a) What pattern does the graph show?
   b) Why is there a pattern?
❸ Estimate the boiling point of the alkane C$_{12}$H$_{26}$.

Activity – How much energy is transferred when a fuel is burned?

Read the following plan for an experiment. It could be used to work out how much heat is produced when a fuel is burned. Then answer the questions that follow.

Plan

- Set up the apparatus as shown in Figure 6.10.
- Put some cold water in the tin held in the clamp.
- Put a lid on the tin, with a hole for a thermometer to go through.
- Put a thermometer through the hole into the cold water.
- Record the temperature of the water.
- Fill the spirit burner with the fuel.
- Weigh the spirit burner and its weight.
- Light the spirit burner. Make sure it has a low flame by adjusting the length of the wick then place it under the tin.
- Put screens round the apparatus to prevent draughts.
- Record the temperature of the water every minute until the water reaches 70 °C.
- Blow out the spirit burner and reweigh it. Record the final weight of the spirit burner.
- Calculate the rise in temperature of the water.

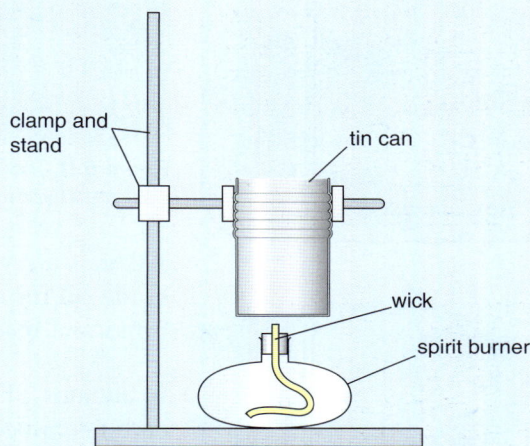

Figure 6.10 Apparatus to find the energy produced when a fuel is burned

You should always use chemicals from an education supplier, rather than commercial products, for this type of investigation. Rooms should be well ventilated for these experiments and you should always wear eye protection.

❶ What important measurement has been missed out of the plan? (Clue: look at line 2 of the plan.)

❷ Some apparatus is not shown in Figure 6.10. Make a list of all the apparatus you would need.

❸ Why do you need screens round the apparatus?

❹ Why do you need a lid on the tin can?

❺ Why is there a hole in the lid of the tin can?

❻ Why do you weigh the spirit burner twice?

❼ One way to present the data for this experiment is in a graph. Draw the graph axes you would use.

What are saturated and unsaturated hydrocarbons?

4 Copy and complete the sentences below, using the words in this box.

> burn   hydrogen
> four   react
> hydrocarbons   fuels

When carbon reacts to form _____ the carbon atoms form _____ bonds. Each _____ atom only forms one bond. Alkanes are very stable and don't _____ much. But they do _____ in air. This is why they are used as _____.

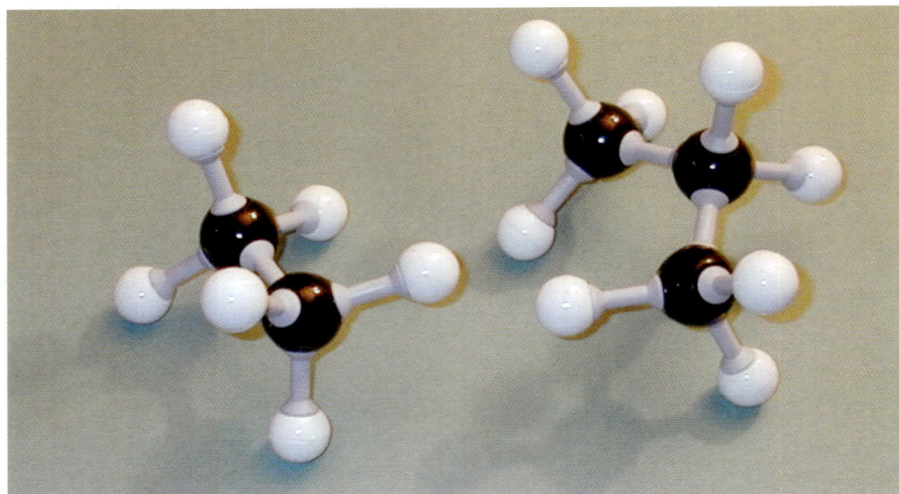

Figure 6.11 The structure of ethane (C_2H_6) and propane (C_3H_8). Ethane and propane are saturated hydrocarbons.

Alkanes are very stable molecules. They have remained unchanged in crude oil for millions of years. They will burn, but that is their only important reaction with other chemicals.

In alkanes, all the carbon atoms are joined together with single carbon–carbon bonds. This means that they form the maximum number of bonds to hydrogen atoms. So they are called **saturated hydrocarbons**.

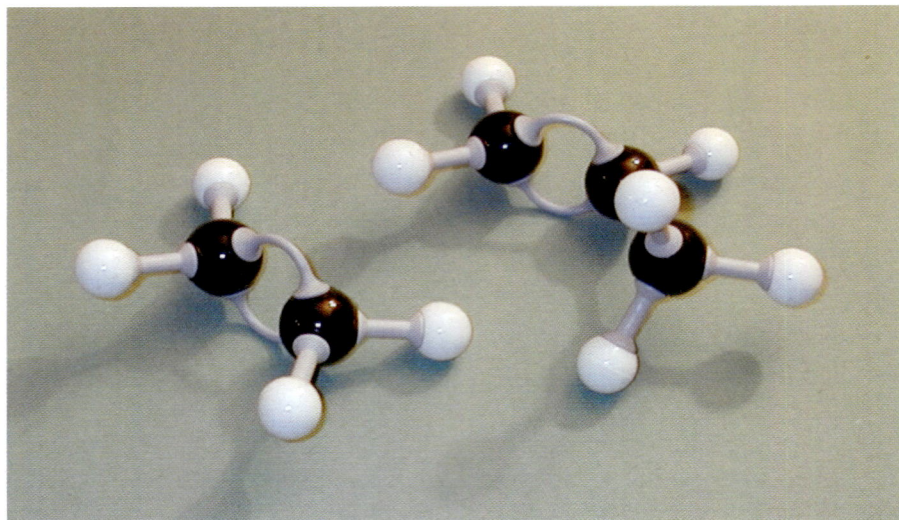

Figure 6.12 The structure of ethene (C_2H_4) and propene (C_3H_6). Ethene and propene are unsaturated hydrocarbons.

Ethene (C_2H_4) and propene (C_3H_6) are very similar to ethane and propane, except that they contain a **double bond** between two of the carbon atoms. These hydrocarbons containing a carbon–carbon double bond could form more bonds to hydrogen atoms. Because of this, they are called **unsaturated hydrocarbons**. The double carbon–carbon bond is very reactive. So unsaturated hydrocarbons have lots of chemical reactions.

6.3 Separating petrol from crude oil

An oil refinery separates crude oil into petrol and other substances by fractional distillation. The fractionating tower in an oil refinery separates the molecules in the crude oil mixture. The biggest molecules are more difficult to vaporise and remain on the lower levels. The smaller molecules vaporise more easily. They move to the higher levels where the temperature is lower. When the temperature is low enough, the smaller molecules condense.

A **fraction** from an oil refinery contains a mixture of compounds with similar boiling points.

The products which collect at the different levels of the fractionating tower are called **fractions**. They are not pure substances, but mixtures of substances with similar boiling points.

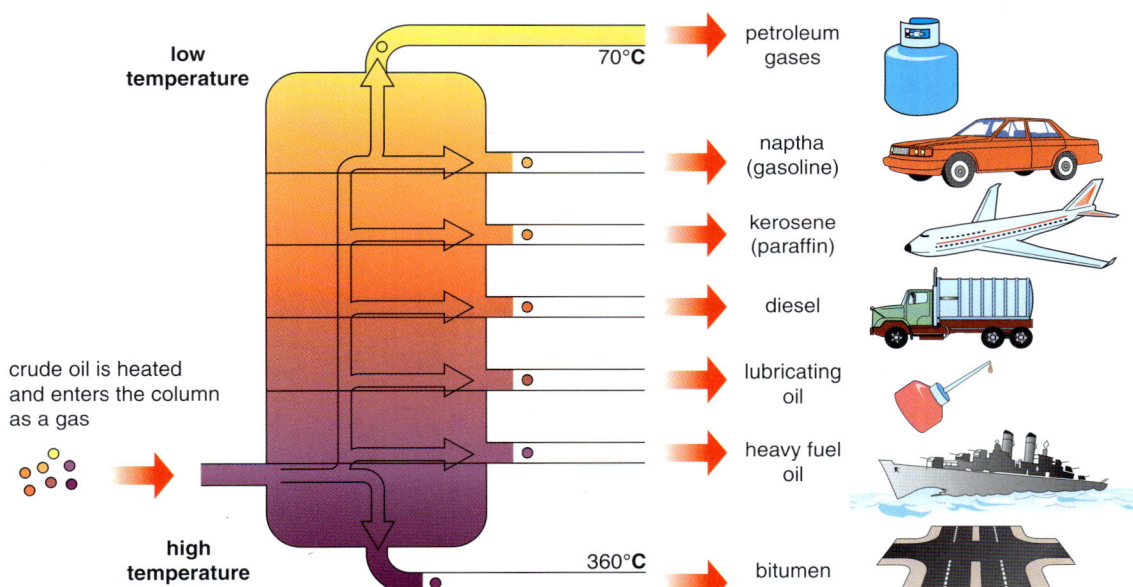

low temperature

70°C — petroleum gases

naptha (gasoline)

kerosene (paraffin)

diesel

crude oil is heated and enters the column as a gas

lubricating oil

heavy fuel oil

high temperature

360°C — bitumen

Figure 6.13 A fractionating tower in an oil refinery

Look closely at Table 6.3 (page 110) as you answer questions 5 to 10.

5 How is the 'lightest' fraction used?

6 What is the naphtha fraction used for?

7 Why can you not get pure octane from the fractionating tower?

8 How does the colour of the fractions change as you move down the fractionating tower?

9 How does the flammability of fractions (how easy it is to burn them) change as you move down the tower?

10 Which fractions are *not* suitable for use as fuels?

Fraction	Description	Flammability	Uses
Petroleum gases	Colourless gases	Explodes if mixed with air and lit	Used as a fuel in the refinery Bottled and sold as LPG
Naphtha	Yellowish liquid flows very easily	Evaporates easily, vapour mixed with air is explosive	Petrol Also used for making other chemicals
Kerosene	Yellowish liquid flows like water	Will burn when heated	Aircraft fuel
Light gas oil	Yellow liquid thicker than water	Needs soaking on a wick or other material to burn	Diesel fuel
Heavy gas oil	Yellow brown liquid	Just burns when soaked onto a wick – very smoky	Used in the catalytic cracker (see Section 6.5)
Lubricant oil	Thick brown liquid like syrup	Needs to be hot and soaked onto a wick before it burns	Grease for lubrication Used in the catalytic cracker
Fuel oil	Thick brown sticky liquid	Needs to be hot and soaked onto a wick before it will burn	Fuel oil for power stations and ships
Bitumen	Black semi-solid	Hardly burns at all unless very hot	Road and roof surfaces

Table 6.3 Fractions from an oil refinery

Activity – Rocville

Rocville is a town on the west coast of Britain. Rocville used to be a seaside resort. But now most people go abroad for their holidays. So there are fewer tourists visiting Rocville. A big oil company wants to build a new oil refinery terminal there.

Here are the views of six people with an interest in the proposed oil refinery. Read carefully what they say.

A town councillor
The town hasn't enough jobs to offer. All our young people are leaving. We need to attract industry to the area, and an oil refinery would be good. There will be jobs involved in building and running the refinery. It will attract other businesses too. The town will benefit.

A fishing boat owner from Rocville
The refinery and its oil tankers will destroy our best fishing grounds. My family have been fishing these waters for two hundred years. The big tankers will pollute the water and kill the fish.

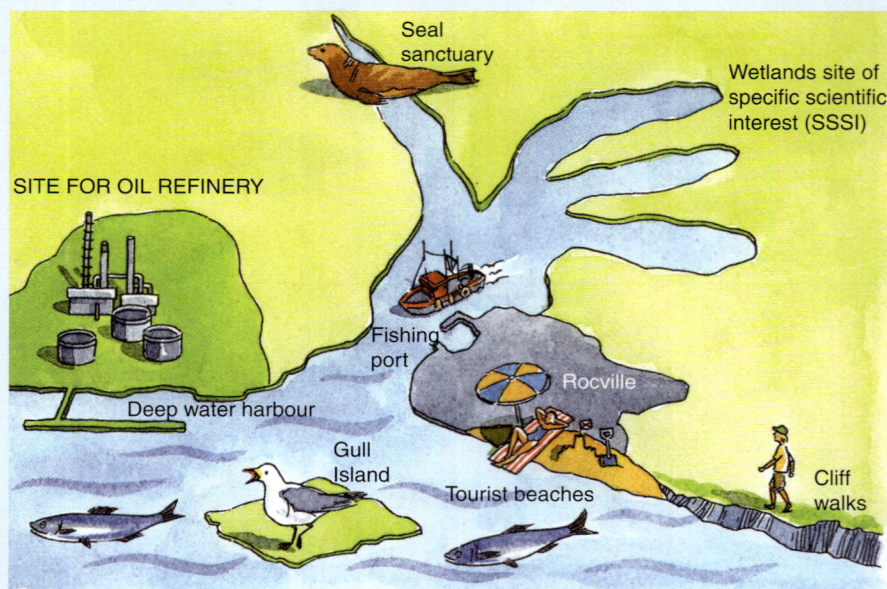

Figure 6.14 Rocville and the surrounding area that could be affected by the new refinery

An executive from the oil company
This site is ideal for us. An oil spill is a risk, but they don't happen often – perhaps once in 60 years. It is true that the refinery will pollute the air, but it will bring people and employment to the area.

An unemployed 17-year old from Rocville
My life is rubbish. The holiday business only has jobs for the summer months. All I want is a regular job and money. I can put up with the smells from an oil refinery if I get that.

A young mother with a 2-month old baby
I've read stories in the paper about dangers from the refinery. Chemicals in crude oil can cause cancer. I visited a refinery. It smelt really foul. I want my child to grow up in a clean, healthy place.

A retired person from Rocville
We retired here after working all our lives. My wife was a classroom assistant and I worked in a sawmill. We want a quiet life and healthy sea air. I have bad asthma. The fumes from the refinery would be the death of me.

1 Write a comment about the proposed refinery that might be made by a young chef in a Rocville restaurant. He doesn't want to see the holiday visitors disappear, but would like to see the town get richer.

2 A landowner in the area has phoned the local radio station about the proposed refinery. Write down what she might say on air. She'd like to make money from selling some land. But she's also worried about her farm workers leaving for higher wages in new jobs.

3 Work in a group of three or four to answer this question.
a) List all the arguments for the refinery.
b) List all the arguments against the refinery.
c) Decide whether your group is in favour of or against the refinery. Explain your decision.

If you have access to ICT, add pictures. You could make a presentation about your decision.

6.4 Problems of burning fossil fuels: air pollution and climate change

Crude oil, and the products made from it, are very useful to people. However, burning fossil fuels releases gases into the air. These pollute the air and cause damage when they form **acid rain**. Scientists also believe these gases cause our climate to change.

Acid rain

Polluted air can produce acid rain. The worst acid rain in the UK fell in Scotland in 1983. The pH value of this acid rain was 1.87. The rain was more acidic than vinegar. It would turn Universal Indicator red.

Rain water is naturally acidic. Rain reacts with carbon dioxide in the air to make weak carbonic acid.

During the last 200 years, the problem has got much worse. Pollution from the burning of fossil fuels has made rain water more acidic. Acid rain attacks marble statues and kills plants. It runs into rivers and lakes where it kills animals and plants in the water.

Figure 6.15 Acid rain can destroy beautiful statues.

Some fossil fuels contain small amounts of sulfur. When these fuels burn, the sulfur reacts with oxygen in the air to form sulfur dioxide. This gas is released (emitted) into the air. Eventually, the sulfur dioxide in the air ends up as sulfuric acid in rain water. Sulfuric acid is a much stronger acid than carbonic acid, so the rain water is much more acidic.

⑪ What colour is Universal Indicator solution in a) pure water b) strong acid?

⑫ What is the acid in unpolluted rain?

⑬ What acids are present in acid rain?

⑭ Copy and complete the sentences below using words from this box.

> creatures   strong   fuels
> statues   kills   sulfuric acid

Acid rain _____ plants. It also kills _____ in water. It attacks marble _____ and ruins them. Acid rain is made when we burn _____. The main polluting substance in acid rain is _____. This is a _____ acid, which spoils the environment.

⑮ Design a poster to explain the effects of acid rain. Use mainly pictures, not words.

Activity – Acid rain

Table 6.4 shows the total worldwide emissions of sulfur dioxide in several years between 1970 and 2003.

Year	1970	1980	1990	1995	2000	2001	2002	2003
Total emissions in thousands of tonnes per year	6500	4800	3700	2400	1200	1100	1000	1000

Table 6.4 Emissions of sulfur dioxide between 1970 and 2003

1 Copy the axes in Figure 6.16 and plot a graph of total emissions against the year.
2 Draw a line of best fit for the points on the graph.
3 An industry spokesperson said:
'The acid rain problem of the 1970s and 1980s is over. There is no longer anything to worry about.'
Do you agree or disagree with what she said? Answer 'Yes' or 'No' and explain your decision.

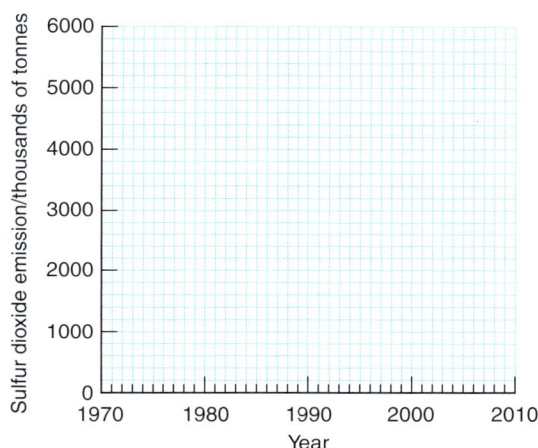

Figure 6.16 Graph to show sulfur dioxide emissions

> **Global dimming** happens when particles in the atmosphere prevent sunlight reaching the Earth. Soot and ash from burning fossil fuels add particles to the air.

Climate change and global warming

The Earth is getting warmer. This is called **global warming**. Scientists think there will be more storms and floods as a result. There is evidence that humans are to blame, by burning fuels.

What is the 'greenhouse effect'?
The 'greenhouse effect' is caused by gases in the air that trap heat from the Sun. Without these gases it would be too cold for life on Earth. Carbon dioxide is one of the 'greenhouse gases'. It is released when we burn fossil fuels.

What is the evidence for global warming?
- Average worldwide (global) temperature increased by 0.6 °C in the twentieth century.
- Sea levels have risen by 10–20 cm.
- Ice in the Arctic Sea has thinned by 40%.

If nothing is done to reduce the emissions of greenhouse gases into the air, scientists predict the following.
- Worldwide temperatures will go up 5 °C by 2100.
- There will be more rainfall overall.
- There will be more droughts in inland areas during hot summers.
- There will be more flooding from storms and rising sea levels.

Other factors may slow down the warming – such as **global dimming**. Scientists are not sure how the balance between global warming and global dimming will play out.

Some people are not so sure
Most people agree that the world is getting warmer. But some people don't believe that global warming is caused by human activity. They say the changes in worldwide temperatures, sea levels and weather patterns are not extraordinary. Similar changes have happened at other times in the Earth's history. And sometimes these changes happened when humans did not exist.

6.5 ## Catalytic cracking

Different fractions of crude oil react in different ways when they are burned. Table 6.5 compares light and heavy fractions from crude oil.

Light fractions	Useful as a fuel?
Petroleum gases Naphtha / petrol Kerosene Diesel fuel	Can be used directly as fuels without any processing
Heavy fractions	
Light and heavy gas oil Lubricants Fuel oil Bitumen	Cannot be used as fuels except in big engines such as those on ships. The oil is heated up before burning

Table 6.5 Light and heavy fractions of crude oil – their usefulness as fuels

Heavy fractions from crude oil contain large hydrocarbon molecules. These large molecules can be broken up by a process called **cracking**. This produces smaller hydrocarbons that are more useful as fuels. Some of these smaller, more useful hydrocarbons are **alkenes**. Alkenes are similar to alkanes but they contain a double bond between carbon atoms.

When heavier fractions are cracked they are vaporised and passed over a catalyst at 500 °C. The catalyst speeds up the reaction. The temperature and pressure are controlled to give smaller hydrocarbons that can be used in petrol. These contain between 5 and 10 carbon atoms. Figure 6.18 on page 114 shows the cracking of a large molecule into several smaller ones.

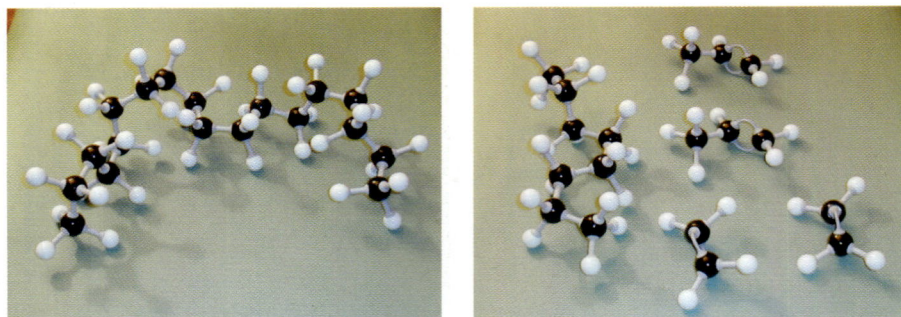

Figure 6.17 Heavier fractions of crude oil produce lots of smoke when they burn. So they are not much use as fuels.

Do you remember?

Unsaturated hydrocarbons contain a double bond between two of their carbon atoms. This means they could form more bonds to hydrogen atoms. Unsaturated hydrocarbons are very reactive.

Figure 6.18 When a large molecule like $C_{18}H_{38}$ is cracked, you can get several fragments. Octane (C_8H_{18}) is one of the molecules found in petrol. Ethene (C_2H_4) and propene (C_3H_6) are good starting materials for making plastics.

Viscosity

Viscosity measures how thick and sticky a liquid is. Thick sticky liquids, such as honey, have a high viscosity. Runny liquids, such as water, have a low viscosity.

Hydrocarbons in the heavy fractions have long, flexible molecules. They are like spaghetti (see Figure 6.19). The longer the molecules, the more tangled they get. The liquids made from these tangled molecules are thick and sticky (**viscous**).

Smaller hydrocarbon molecules, like those in naptha (petrol), are much shorter. They are more like macaroni (see Figure 6.19). The short molecules don't get tangled. The liquids made from these molecules are runny and easy to pour.

Figure 6.19 Which of these pasta shapes becomes more tangled?

16 Name two fuels you can get from crude oil without the 'cracking' process.

17 Why are the heavier fractions less useful as fuels? (Hint: look at Figure 6.17, page 114.)

18 Suggest another name for viscosity.

19 Copy and complete the sentences below using words from this box.

cracking	petrol	shorter	useful
long	double	use	broken
	hydrocarbon	crude	

A lot of _____ oil is made up of large _____ molecules that are of little _____ except as engine oil. These long molecules are _____ up into _____ molecules that are more useful. The long molecules are turned into shorter molecules of _____ and diesel fuel. These are very _____ materials. The process of breaking up the large molecules is called _____. Some of the shorter molecules produced have _____ bonds between carbon atoms.

6.6 # Alkenes

In Section 6.2, you saw that the alkanes form a series of similar hydrocarbons. The **alkenes** also form their own series of similar hydrocarbons.

Each alkene has *one* carbon–carbon *double* bond in its molecule. So, the simplest alkene is ethene (C_2H_4) (Figure 6.21). If you add one $-CH_2-$ link to the molecule of ethene, it becomes propene (C_3H_6). This is the next molecule in the series of alkenes (Figure 6.22).

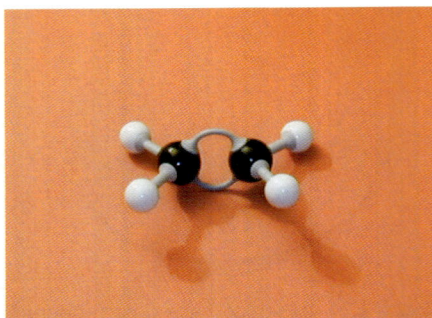

Figure 6.20 The molecular model of ethene

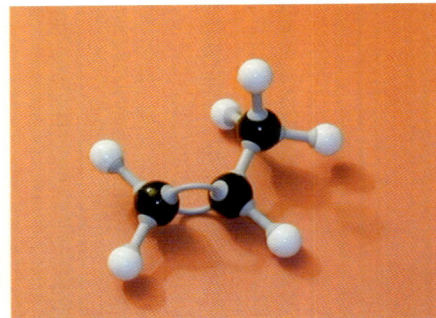

Figure 6.21 The molecular model of propene

> **20** Copy the model of a propene molecule in Figure 6.21 and label the following parts:
> - A hydrogen atom
> - A carbon atom
> - A carbon-to-carbon single bond
> - A carbon-to-carbon double bond

By adding a $-CH_2-$ link each time, we get the next three members of the series: butene (C_4H_8), pentene (C_5H_{10}) and hexene (C_6H_{12}).

The reactivity of alkenes

The carbon–carbon double bond makes alkenes very reactive. The double bond can open up and react, but the skeleton of the molecule stays the same (Figure 6.22).

For example, alkenes will react easily with bromine water to form a new substance and remove the orange colour of bromine. This is used as a test to see if a substance contains a double carbon–carbon bond (see page 135).

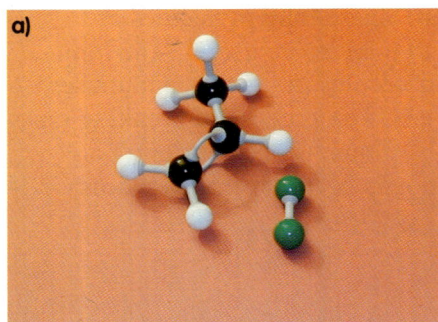

Figure 6.22 When propene and bromine water (green) react, one of the bonds in the double bond of the propene molecule opens up.

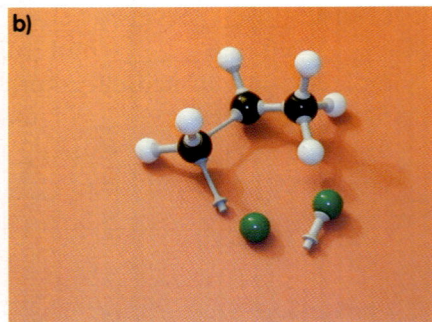

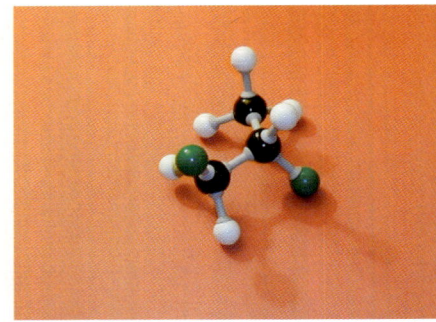

Figure 6.23 After the double bond opens up, each bromine atom joins a carbon atom.

6.7 Ethanol – spirit fuel

Brazil has started to sell 'spirit motor fuel' based on ethanol made from sugar cane

Figure 6.24 In this photo, a piece of cotton wool has been soaked in ethanol and set alight. Ethanol is a clean, safe fuel.

Ethanol has a lot of advantages as a fuel.
- It burns with a clean, smokeless flame.
- It is not oily and can be washed away with water.
- It mixes with water and does not pollute it.
- It occurs naturally (for example, in rotting fruit).
- Cars can run on it without needing to have different engines.

Making ethanol from ethene

You can make ethanol from ethene and steam (water). The atoms from a water molecule are added to the double bond in ethene (Figure 6.25).

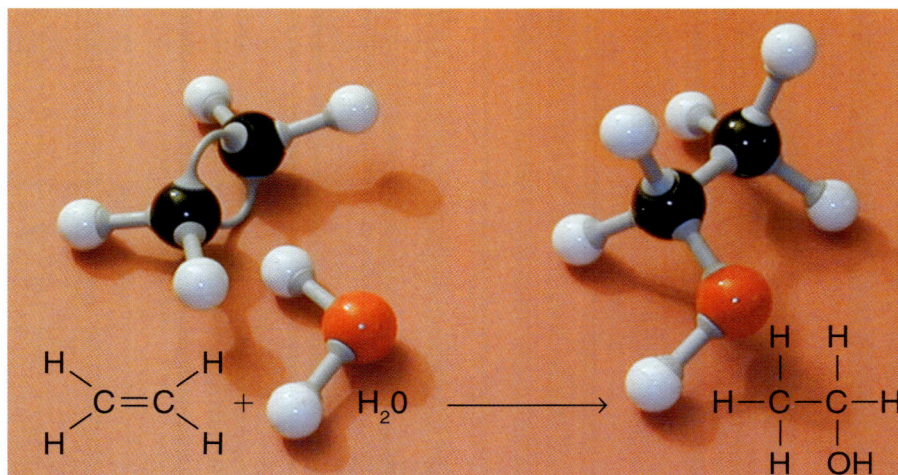

Figure 6.25 Making ethanol from ethene and steam

In a reactor, hot ethene and steam are mixed under pressure. Even with all the molecules pushed together and colliding with each other, the reaction is very slow. But a catalyst speeds up the reaction.

Making ethanol by fermentation of plant products

Starchy material can be used to make ethanol. The process is called **fermentation**.

Any starchy material such as barley or sugar cane can be used. The process is as follows.
- Heat the plant material with water. Natural enzymes in the plant material break up the starch molecules into smaller sugar molecules.
- Yeast in the plant material (or added separately) changes the sugars into ethanol and carbon dioxide. The carbon dioxide bubbles off as a gas.
- Keep the mixture warm for several days until fermentation is complete.
- Separate ethanol from the watery mixture by **fractional distillation**.

The word equation for the fermentation process is:

$$\text{sugar} \xrightarrow[\text{enzyme}]{\text{yeast}} \text{ethanol} + \text{carbon dioxide}$$

㉑ Copy and complete the table below to explain why ethanol is a good fuel.

	Why ethanol is a good fuel
a) The flame you get	
b) When it gets into water	
c) When it is used in cars	

㉒ Copy and complete the flow chart in Figure 6.26 to show how ethanol is produced by fermentation.

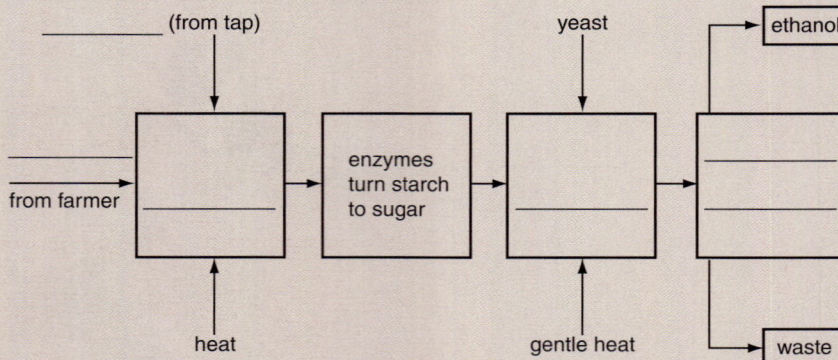

Figure 6.26 A flow chart to show how ethanol is produced by fermentation

6.8 Polymers

A **monomer** is a small molecule that makes links with other molecules of the same kind to form one long molecule. This long molecule is called a **polymer**.

(Mono means 'one'. Poly means 'many'.)

A polymer is made from lots of monomers joined together.

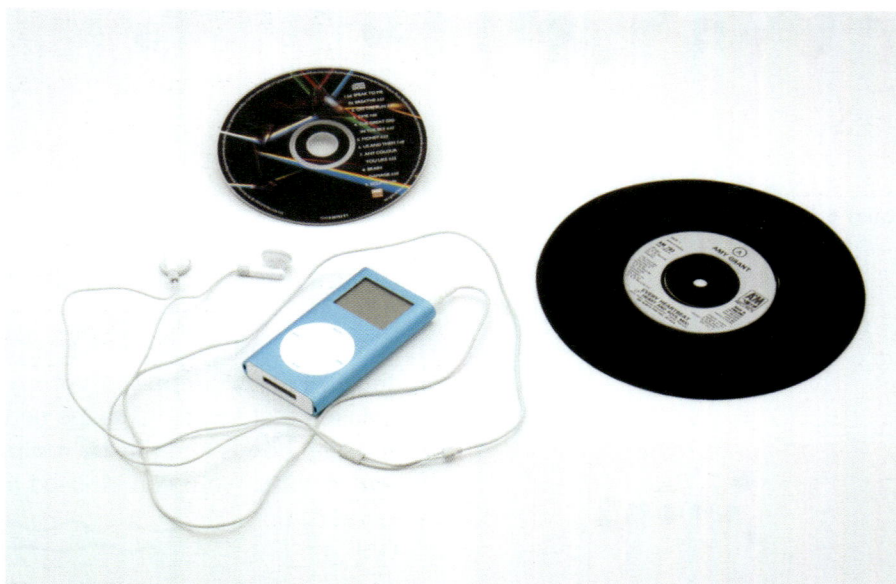

Figure 6.27 About 20 years ago, records started to be replaced by CDs. Records and CDs are made of 'vinyl', or PVC. PVC is a manufactured plastic. Without plastics, we wouldn't be able to enjoy recorded music.

Plastics are materials called **polymers**. They are made from various alkenes. Different plastics have different properties. They all have the useful property that they can be moulded or shaped into different forms.

Polymers are made up of molecules with long chains of atoms.

Figure 6.28 Bags, and other items made of polythene, litter our streets and many beaches. They get into rivers and the sea after being thrown away. They remain there for years and years.

Polythene is the plastic used for supermarket carrier bags and for most packaging material. It is light and can be thin and flexible. The problem with polythene is that it is not **biodegradable**. It does not rot away like wood or paper.

Polythene is made from ethene. The double bonds in ethene molecules open up and link the molecules together. Thousands of ethene molecules can be linked in this way, creating a very long molecule of polythene.

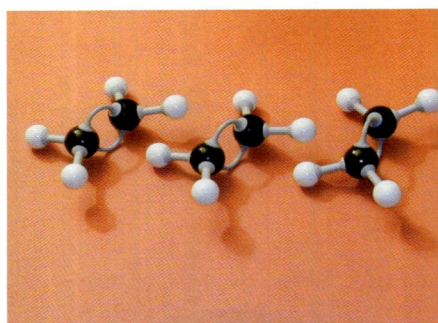

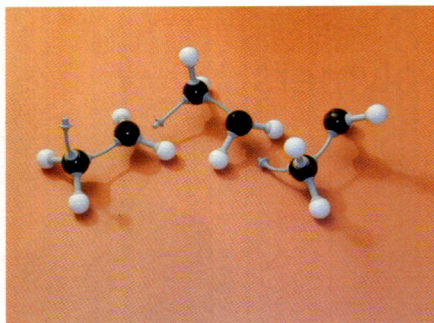

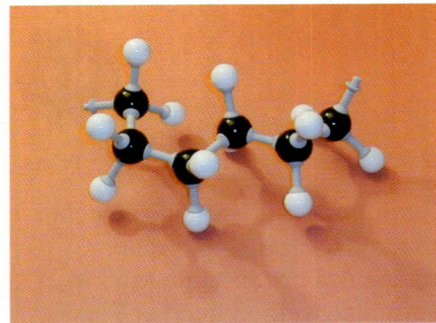

Figure 6.29 How polythene molecules are formed from ethene

Choosing the monomer to give the best plastic

The properties of the monomer affect the properties of the polymer produced. **Polypropene** (polypropylene) is made from propene. It is a stiff plastic that keeps its shape better than polythene. It is used for washing up bowls and plastic toolboxes. Although it is stiff, thin pieces are still flexible, so it can be used to make its own hinge.

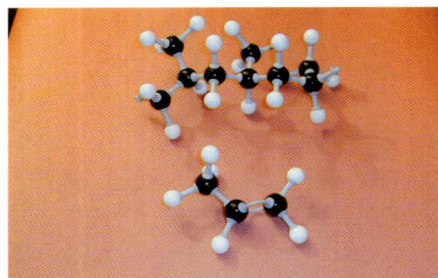

Figure 6.30 Propene and polypropene

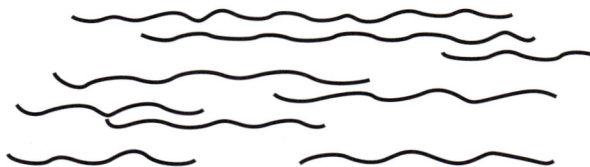

flexible polymer molecules

crosslinks make the structure more rigid

Figure 6.31 Some polymers are very hard and rigid. This is because links (bonds) have formed between the longer chains as monomers join together. These **crosslinks** stop the polymer molecules from being flexible.

㉓ Explain the difference between a monomer and a polymer.

㉔ Copy and complete the sentences below using the words in this box.

flexible	stiffer	plastics
rot	properties	invented
not	polymers	alkenes

Records, CDs and DVDs are all things we would _____ have if plastics had not been _____. Plastics are materials called _____. They can be made from _____ Plastics can be very _____ so they are used as packaging. The _____ of the monomer affect the properties of the _____ produced. For example, some plastics are _____ than others. The main problem with plastics is that they do not _____ away like many other materials.

㉕ Copy Figure 6.32 below. It is a simple model for a crosslinked plastic. Label the following parts on your diagram.
a) a monomer unit
b) a polymer chain
c) a crosslink

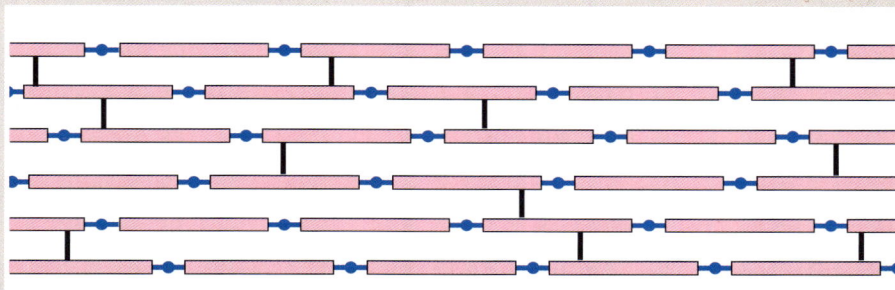

Figure 6.32 Crosslinked plastic

6.9 Recycling plastics

Plastics have lots of uses because of their desirable properties.
- They are light in weight.
- They are very long lasting.
- They are acid resistant.
- They are not affected by water and are strong when wet.
- They are tough and resist knocks.
- They are good insulators of heat and electricity.
- They are relatively cheap to produce.

The problem with plastics

Most plastics do not rot away like other materials. They are not biodegradable. There are no animals or plants that eat the plastics. So plastics take a long time to break down. More and more plastics are being dumped with rubbish in landfill sites. But this uses up valuable space. At the moment only 7% of plastic waste is recycled. We need to find ways to recycle more plastic. Re-using plastics is even better than recycling. It uses less energy and reduces waste.

Some supermarkets and shops use re-usable and returnable plastic crates for transporting and displaying goods.

Making biodegradable plastics

Biodegradable plastic carrier bags have been produced in the last few years. These bags are made from a plastic which rots away, but not on the journey home!

Some of the biodegradable bags are made from plant-based monomers, which rot and break up the plastic molecules. Others decay when exposed to sunlight for a long time. In Sweden, McDonalds are now using biodegradable cutlery. This means that all their catering waste can be turned into compost.

But these new biodegradable plastics are still a worry.
- The plastics need special conditions to rot.
- The plastics produce pollution while they rot.
- People may use and throw away more of these plastics if they believe they will rot.

---

26 Look carefully at the pie chart in Figure 6.33.
a) What percentage of your rubbish is plastic?
b) What rubbish could you put on a compost heap?
c) What percentage of the total rubbish is this compostible rubbish?

27 We recycle paper cardboard, glass and metal. What percentage of the total rubbish is this recycled material?

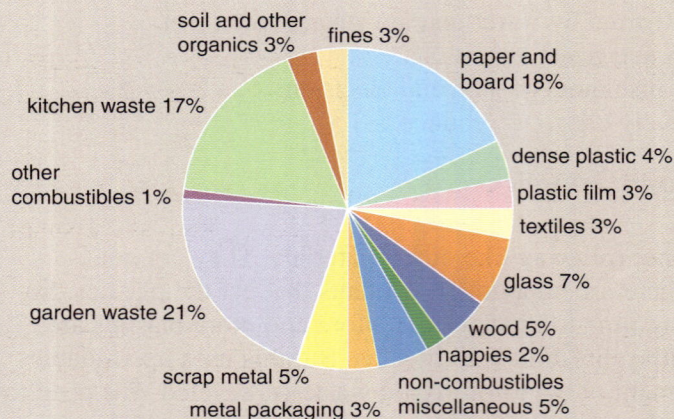

28 Look carefully at the uses of plastics in Figure 6.34.
a) What percentage of plastic is used for packaging in the UK?
b) Give two examples of plastic used in building your school.

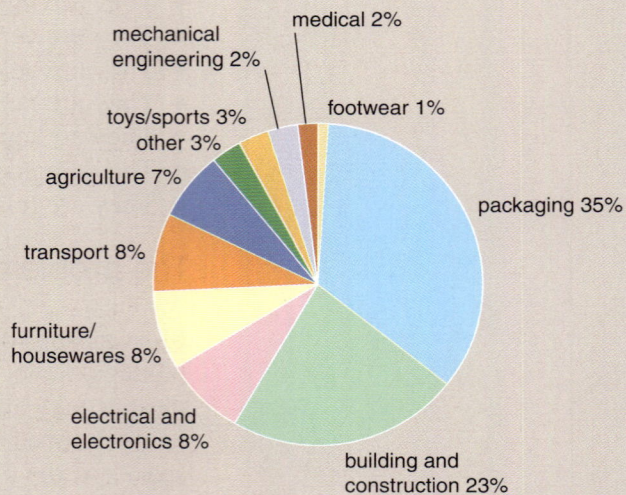

soil and other organics 3%
fines 3%
kitchen waste 17%
paper and board 18%
other combustibles 1%
dense plastic 4%
plastic film 3%
textiles 3%
glass 7%
garden waste 21%
wood 5%
nappies 2%
scrap metal 5%
non-combustibles
metal packaging 3%
miscellaneous 5%

Figure 6.33 The rubbish you produce

mechanical engineering 2%
medical 2%
toys/sports 3%
footwear 1%
other 3%
agriculture 7%
packaging 35%
transport 8%
furniture/housewares 8%
electrical and electronics 8%
building and construction 23%

Figure 6.34 The uses of plastics in the UK

29 Read the following comment and then answer the question.

Do you agree or disagree with this man? Say 'Yes' or 'No' and explain your answer.

Plastic is just oil on its way to being burned. Instead of just burning oil as a fuel, the oil gets a new life as a plastic bag. Then it's burned as a fuel. The energy released can be used to generate electricity, or for heating. People don't have time to sort household waste and send it to the right place for recycling. Life in the twenty-first century is just too busy for people to do that.

Figure 6.35

Producing 'smart' polymer materials

There are many different uses for plastics. Companies spend millions of pounds in developing new plastic materials. 'Smart' materials are designed to do very special jobs. They are used in medicine, industry and in our homes.

Shape memory polymers

Shape memory polymers 'remember' the shape they were made in. Even when they are bent and changed, they can spring back into their original shape.

Doctors can put a straightened plastic thread into a patient's artery. The shape memory then makes the thread return to being a coiled spring. This widens the artery allowing the patient's blood to flow better.

A surgeon uses stitches made of shape memory polymers to hold the flesh together. Their memory shape makes them slowly tighten, making a better repair.

In dentistry, teeth braces can be made in the shape of the mouth. They are then hardened by using a special lamp. (There is more about smart metals in Section 5.7.)

New dressings containing hydrogel, which is a smart polymer, will react to the state of a wound. Sometimes they keep it moist. At other times they dry it out. They are useful for messy wounds.

Contact lenses

Ordinary contact lenses are easily scratched and can be difficult to keep clean. There are now contact lenses that can be worn once and thrown away.

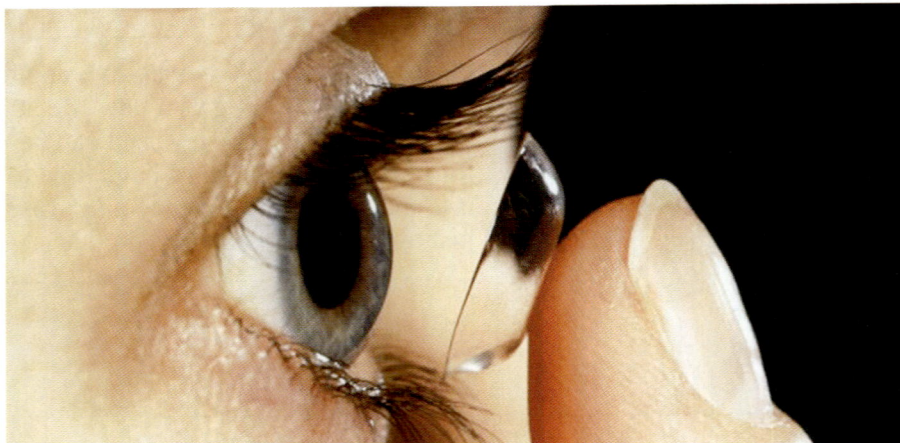

Figure 6.36 Improvements in properties of polymers have allowed us to make throw-away contact lenses.

Mobile phone takes itself apart

If a mobile phone like one of those in Figure 6.37 is heated, the shape memory polymers make it fall apart into pieces. The expensive materials can then be sorted and recycled. This also makes it easier to get rid of any parts containing poisonous chemicals.

Figure 6.37 Smart materials allow the phone to fall apart for easy recycling.

Supergel

Supergel is a smart polymer. It can absorb lots of oil and is very useful in dealing with oil slicks on the sea. The gel absorbs the oil and thickens. The resulting mixture sets so solid that it can be rolled up and lifted out of the seawater. This helps to clean up the pollution after an oil tanker has leaked.

Figure 6.38 Supergel for cleaning oil slicks

Conducting polymers

Conducting polymers are amazing materials. Electricity makes them contract, stretch or twist like a muscle fibre. Incredibly small nano-robots of the future will use these as moving parts.

30 Copy and complete the sentences below using words from this box.

> smart   stitches   recycling   arteries   shape
> special   oil   phone

Smart materials have been designed to have _____ uses. For example _____ memory plastics can be used to hold _____ open to let the blood flow better. They can be used for _____ that hold wounds closed. Supergel is a _____ polymer. It can be used to absorb _____ from the surface of the sea. The tiny bits inside a mobile _____ can be separated for _____ if the phone is put together using a smart material that falls apart when it is heated.

Summary

✓ **Hydrocarbons** are compounds that contain atoms of hydrogen and carbon only.

✓ **Crude oil** is a mixture of hydrocarbons which can be separated into **fractions** by **fractional distillation**.

✓ The **alkanes** are a series of similar **saturated hydrocarbons**. Alkanes are found in crude oil and are mainly used as fuels.

✓ Saturated hydrocarbons are stable. **Unsaturated hydrocarbons** are reactive.

✓ Burning fossil fuels is a major cause of air pollution and global warming.

✓ The heavier fractions from crude oil can be changed into useful hydrocarbons by a process called **cracking**. They can then be used to produce petrol.

✓ The **alkenes** have a carbon–carbon double bond which makes them more reactive than alkanes.

✓ Plastics are examples of **polymers**. They can be made from alkenes.

✓ Polymers are large molecules made by joining together smaller molecules called **monomers**.

✓ Ethene can be polymerised to form polythene.

✓ Different plastics can be made, with a wide range of properties and uses.

✓ Plastics create waste problems because most of them are not recycled and they do not rot away.

EXAM QUESTIONS

❶ Copy and complete the table below about fuels.

Fuel	Where the fuel is used	Where it comes from
Methane (natural gas)		In the ground
	Motor cars	From crude oil
Diesel		

(4 marks)

❷ a) What is crude oil?
b) Where is it found?
c) What is a hydrocarbon? *(3 marks)*

❸ a) Explain the difference between a saturated hydrocarbon and an unsaturated hydrocarbon.
b) Give an example of
(i) a saturated hydrocarbon;
(ii) an unsaturated hydrocarbon. *(3 marks)*

❹ How would you test a hydrocarbon to see if it is unsaturated? *(3 marks)*

❺ a) What is acid rain? *(1 mark)*

b) Which acid is mainly responsible for acid rain? *(1 mark)*
c) Explain how this acid gets into the environment. *(1 mark)*
d) Explain the effect of acid rain on marble statues. Give as much detail as you can. *(2 marks)*
e) Write a word equation for the reaction that produces acid rain. *(2 marks)*
f) Describe the effect of acid rain on:
i) trees;   ii) fish. *(2 marks)*

❻ Some fractions of crude oil were tested for viscosity using the apparatus in Figure 6.40.
a) Explain how you could use this apparatus to compare the viscosities of fuels. *(2 marks)*
b) Name one factor that must be constant to make this a fair test. *(1 mark)*
c) The results of four tests are shown in the table on page 127. Copy the graph ages shown in Figure 6.41 and plot the results from the table onto your graph. Make your graph a bar chart or a line graph, whichever you think is more suitable. *(2 marks)*
d) Explain the pattern in the results. (Hint: Use what you know about the different molecular sizes in the fractions.) *(2 marks)*

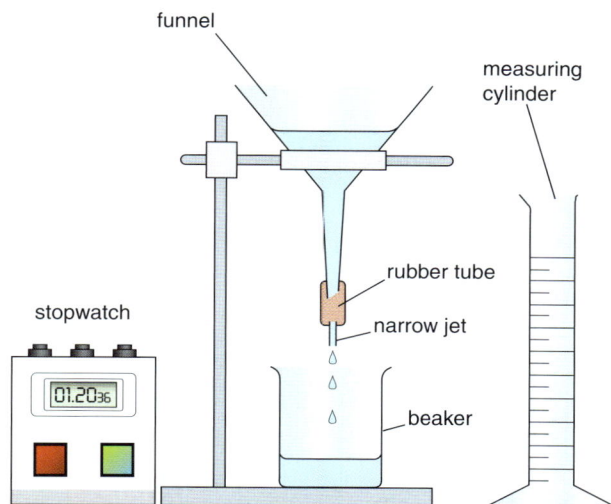

funnel

measuring cylinder

stopwatch

01.20₃₆

rubber tube

narrow jet

beaker

Figure 6.40

Fraction	Time in minutes
Petrol	2.0
Kerosene	2.3
Diesel	3.0
Lubricating oil	6.0

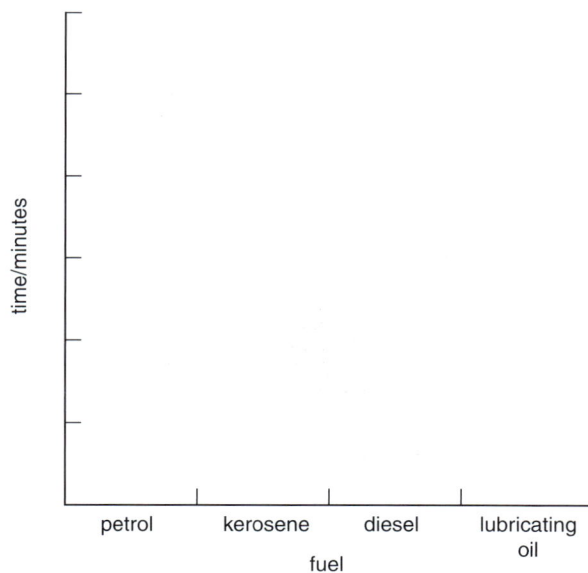

time/minutes

petrol   kerosene   diesel   lubricating oil

fuel

Figure 6.41

7 Crude oil is separated by fractional distillation using A column as in Figure 6.42.
a) How does fractional distillation work?
(2 marks)
b) Why does petrol burn more easily than diesel oil? *(2 marks)*
c) Petrol contains a saturated hydrocarbon with the formula C_7H_{16}. Draw the structural formula of this compound. *(2 marks)*
d) Lubricating oil is used in a catalytic cracker. Describe what a catalytic cracker does. *(2 marks)*
e) Draw three molecules that you might obtain from cracking one molecule of C_1H_{32}. *(2 marks)*

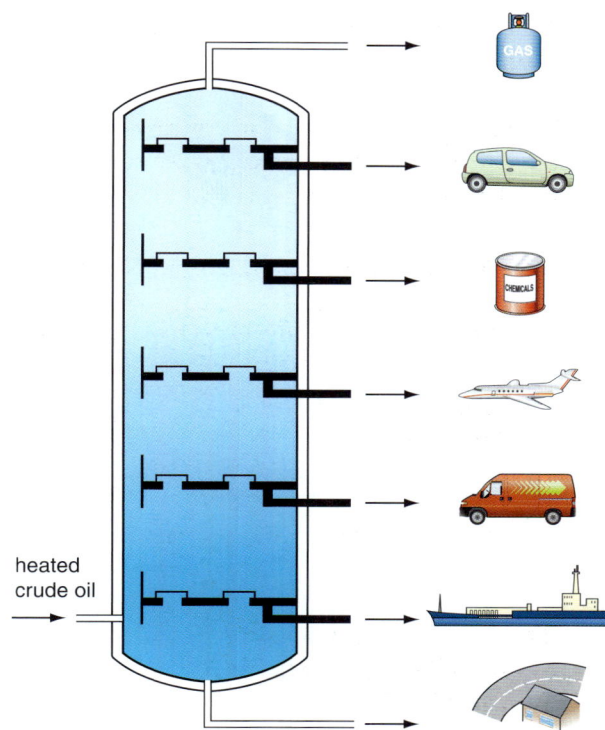

heated crude oil

Figure 6.42

EXAMQUESTIONS

Chapter 7
How can plant oils be used?

At the end of this chapter you should:

- ✓ know that plants provide us with plant oils;
- ✓ know that natural products from plants can be changed chemically to produce new substances;
- ✓ understand how we get plant oils from plants;
- ✓ know that plant oils are used to make foods;
- ✓ know that plant oils are used to make biodiesel fuel;
- ✓ know about the additives in processed foods.

Figure 7.1 Plants are like chemical factories. They take in raw materials and turn them into materials for us to eat or use.

Plant products

Plants contain proteins, carbohydrates, fats, vitamins, fibre and minerals. All these are needed for a balanced diet.

A plant collects energy during photosynthesis. This makes starch in the leaves of the plant. Plants also store energy as oil in their seeds. This energy helps the seed to grow.

Many plants contain oils. Plant oils (often called vegetable oils) have been used for thousands of years. They are used in foods for cooking oil, in cosmetics and as lamp oil. Years ago, the oil was extracted by squeezing the plant in a press until the oil flowed out. More modern methods now extract the oil by dissolving it in a solvent. The solvent is then distilled off leaving the oil behind.

Figure 7.2 Plants capture energy from sunlight that ends up in our food.

❶ How do plants take in water?

❷ How do plants take in carbon dioxide?

❸ Name two products obtained from plant oil.

7.2 # Oil in the kitchen

❹ a) What food groups are essential for a healthy balanced diet?
  b) Are there any food groups we cannot get from plants?

❺ Old fashioned oil lamps burned plant oil. Where did the energy come from that was transferred as light when the oil burned?

Plant oil	Where the oil is stored in the plant
Olive oil	Fruit
Rape oil	Seeds
Peanut oil	Nut (food store for seeds)
Avocado oil	Fruit
Sunflower oil	Seed
Palm oil	Fruit

Table 7.1 Some common plant oils and the sources of the oil

Oil and water do not mix. Shake a container of oil and water for as long as you like. As soon as you stop, the oil and water will separate. The water will sink to the bottom and the oil will float to the top.

But oil and water can be made to mix. To do this you need an **emulsifier**: a substance that enables oil molecules to link to water molecules. Emulsifier molecules have a water-loving end and a water-hating end. So emulsifiers can bring oil and water together and allow them to mix. The mixture of oil and water is called an **emulsion**. Egg yolk, mustard and sugars can act as emulsifiers.

Figure 7.3 'Plant oils provide nutrients and lots of energy. But eating too much food that is very high in energy will make you fat!'

Figure 7.4 Droplets of water on a waxed surface. Oil and water do not mix.

Figure 7.5 Soap is a common emulsifier.

Figure 7.6 Salad dressings such as vinaigrette and mayonnaise mix oil and water. The thick mixture coats the salad leaves.

To make an emulsion of oil in water, the oil is broken up into tiny droplets. Shaking or whisking breaks the oil into droplets. The emulsifier is then used to prevent the droplets from pooling together again.

Emulsions can be temporary like vinaigrette, or permanent like mayonnaise. The thicker an emulsion is, the less likely it is to separate because the droplets move more slowly through a thick mixture.

Fresh milk is an emulsion of tiny fat droplets in water. Ice cream is also an emulsion. Tiny crystals of flavoured ice are trapped in droplets of cream or vegetable fat. Tiny bubbles of air are also whisked into the mixture.

Figure 7.7 Mayonnaise was invented in 1756 by a French chef. He made a feast after a French victory over the British at Mahon in Minorca. The chef planned to make a sauce using cream and eggs. But, as there was no cream, he invented a new sauce made of olive oil and eggs. He called it 'mahonnaise'.

Figure 7.8 There are many uses for emulsions. Ice cream has tiny ice crystals suspended in oily droplets. This gives it a smooth, creamy texture. Many emulsions used by chefs are sauces. They may be sour, fruity or spicy. Gravy and salad dressings are thick so that they stick to the food.

Frying in oil

Frying is a fast method of cooking. Plant oils can be heated to 230 °C before they start smoking, but water can only be heated to 100 °C. So fried food cooks faster. You should always fry with oil above 100 °C to seal in the flavour. Above 100 °C, steam escapes from the food and prevents oil soaking in. Frying also adds flavour. The high temperature changes the surface of the food so it turns brown. This makes it taste good. You should always take care when helping to prepare fried food. In the UK alone there are 300 injuries from chip pan fires every year!

Figure 7.9 Frying adds to the flavour.

6 Why are sauces and salad dressings usually emulsions?

7 Why do chips cook faster than boiled potatoes?

8 Why do chips taste different from boiled potatoes?

9 Copy and complete the sentences on the right using words from the box.

gravy	salad	ice-cream	separating
coat	fruit	emulsion	kitchen
	sunflower	emulsifier	

Plant oils have lots of uses in the _____. Plant oils such as olive oil come from _____. Plant oils such as _____ oil come from seeds. Plant oils are often made into _____ dressings, sauces and _____. These have to be thick to _____ the food. To make a sauce you mix oil and a watery liquid. The _____ stops the oil and water from _____. The mixture is called an _____. Mayonnaise is an emulsion. _____ is an emulsion with ice crystals in it.

7.3 Food additives

Additives are used to make food look and taste better. They also stop food going stale. But some people are concerned that food additives can make unhealthy foods look more attractive than healthy, fresh foods.

Adding chemicals to foods isn't a recent invention. Nitrates have been used since the Middle Ages to preserve meat. Laws passed by the European Union (EU) require food additives such as nitrates to have an E-number. If an additive has an **E-number**, it means it has passed safety tests. All food labels must list the additives contained in the food.

Figure 7.10 Lots of foods contain additives.

What are the different types of additives?

The most common additives are:
- antioxidants which keep food fresh;
- food colourings;
- emulsifiers, stabilisers and thickeners;
- flavourings;
- preservatives;
- sweeteners.

Additives have some benefits.
- Antioxidants stop butter going off when it's left out of the fridge.
- Additives prevent bread going mouldy.
- Preservatives kill the bacteria that produce deadly poisons in meats.
- Ten tiny sweeteners have the same sweetening effect as a kilogram of sugar.
- Sweeteners are 'low calorie' and harmless to teeth.

Many additives are natural substances. All the substances listed below are in *fresh home-grown* tomatoes. They have not been added. These substances have E-numbers but they are in the tomatoes naturally. Sometimes these substances are added to other foods to improve them.

Flavour enhancer
- E621, monosodium glutamate

Colours
- E160a, carotene
- E 101 (vitamin B6)

Antioxidant
- E300 (vitamin C)

Acids
- E330, citric acid
- E296, malic acid

Additives can also have drawbacks.
- Aspartame, a sweetener, is used in soft drinks. When aspartame was fed to rats, it caused violent behaviour.
- Tartrazine (E102), the orange colour in some soft drinks, has been linked to asthma and hyperactivity.
- Sunset Yellow (E110), used in biscuits, has been found to damage the kidneys.

Figure 7.11 Tomatoes contain several natural substances that are used as 'additives' in other foods.

Figure 7.12 Analysing food additives by chromatography

Analysing food additives

Chemists use chromatography to test foods for artificial colours. This is the same idea as separating ink colours with water on filter paper.

10 Why are additives added to some foods?

11 Additives are given E-numbers. Why are these numbers used?

12 a) Write down three benefits of some food additives.
   b) Write down three drawbacks of some food additives.
   c) If you were a young parent, would you give your child food containing additives? What is your opinion on this?

13 Copy and complete the sentences below using words from this box.

labels	food	good	unhealthy	natural	additives
		tomatoes	poor		

Food _____ can stop food from going bad. But food additives can also make _____ quality food look nice and taste _____. So _____ additives can make your diet _____. All food _____ must list the additives contained in the food. Some additives are _____ substances and can be found in fresh foods such as _____.

Figure 7.13 Sunflowers are a common crop in hot countries.

7.4 The marge story

You will have seen pictures of sunflowers on margarine tubs. But, how do we get margarine from sunflowers?

Sunflower oil is extracted from sunflower seeds. The seeds are crushed between heavy rollers and the oil is squeezed out.

Olive oil is also used to make margarine. There are several grades of olive oil. Extra virgin olive oil is produced by pressing ripe olives at room temperature. It is the highest quality of olive oil. Growers can use special machines to get more oil out of the olives. But this oil is acidic and used for soap making.

Butter is made from the cream in milk. This is an animal fat which contains cholesterol. Too much cholesterol causes heart disease.

Margarine is a butter substitute. You can remove a lot of the cholesterol from your diet by eating margarine instead of butter.

How do we make solid margarine from a liquid plant oil, such as sunflower oil?

- Plant oils contain **unsaturated** molecules with a carbon–carbon double bond (see Section 6.2).
- If you add hydrogen to the double bond, the oil turns into a much more solid material – much easier to spread on bread.
- This oil is heated to 60 °C with a powdered nickel catalyst and hydrogen gas is bubbled through it.
- After the reaction, the oil is filtered to remove the nickel and allowed to cool.
- As the liquid cools, it forms solid margarine.

⓮ Why could eating a lot of olives make you fat?

⓯ Why is extra virgin olive oil the best for cooking?

⓰ Design a simple poster to tell people the difference between butter and margarine.

⓱ Copy and complete the sentences on the right using words from the box.

| cools | nickel | unsaturated | added |
| cholesterol | sunflower | butter |
| hydrogen |

Margarine is healthier than _____ because it contains less cholesterol. Too much _____ can cause heart disease. Margarine is made from plant oils such as _____ oil. The oil is heated at 60 °C with a _____ catalyst and _____ is bubbled into the mixture. During the reaction, hydrogen is _____ to the _____ molecules of oil. The mixture forms a solid when it _____.

7.5 ## Polyunsaturated fat from plant oils

An unsaturated fat has hydrocarbon chains with some carbon–carbon double bonds.

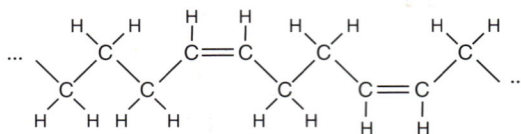

A saturated fat has hydrocarbon chains in which all carbon–carbon bonds are single.

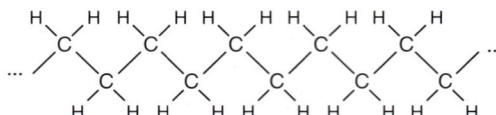

Figure 7.14 The structure of molecules of saturated and unsaturated fats

Saturated fats are found in meat, milk, cheese and cream. If we reduce the amount of saturated fat in our diet, we also reduce our chances of developing heart disease. Because of this, it is important to read food labels and check how much saturated fat there is in the food.

Unsaturated fats are found in most plant oils. Choosing foods containing unsaturated fats is healthier because they will reduce your risk of heart disease (see also page 22).

But fat is fattening and unsaturated fats, like saturated fats, have lots of energy (calories). So eating too much unsaturated fat will increase your weight just as saturated fat will.

Here are seven tips to help you reduce fat in your diet.
- Eat less fatty food.
- Eat more low-fat foods such as fruit, vegetables, bread, rice, pasta and cereals.
- Use less fat, less oil, less butter and less margarine in cooking.
- Use skimmed milk and low-fat cheeses instead of whole milk and cheese.
- Choose margarine instead of butter.
- Trim fat from meat and remove the skin from chicken.
- Read and compare food labels to find foods that have less fat.

⓲ Why is 'saturated' fat bad for you?

⓳ Why is eating too much unsaturated fat also bad for you?

⓴ How would you test olive oil to show that it contained unsaturated fats?

㉑ Copy and complete the sentences below using words from this box.

> eat   healthier   diet
> plant   disease   saturated
> cheese   bromine
> unsaturated

Meat, milk, cream and _____ contain saturated fats. _____ fats can cause heart _____. Unsaturated fats are found in _____ oils. These are much _____ as part of your diet. You should try to _____ low-fat food. You should also make sure more fat in your _____ is from plant oils rather than saturated fats. _____ water is used to test whether a fat is unsaturated. If the colour fades rapidly, it means that the fat is _____.

Activity – Testing an oil or fat to see if it is unsaturated

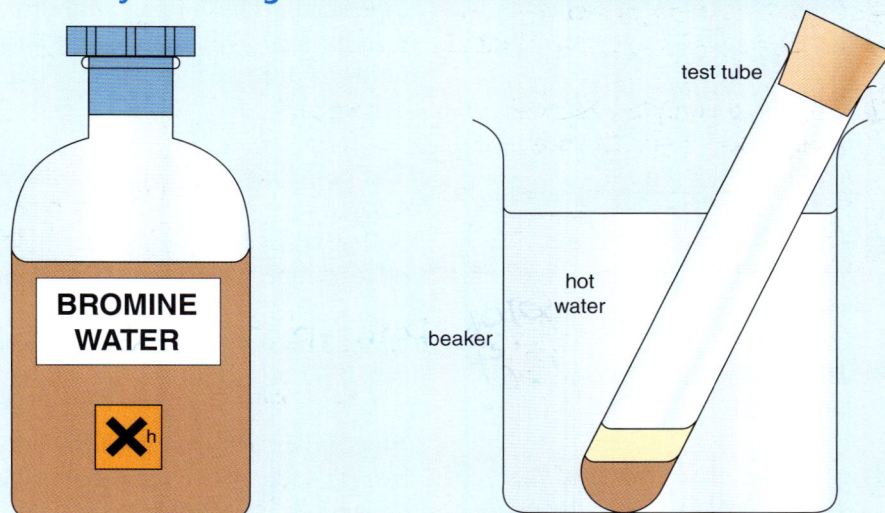

Figure 7.15 Testing for unsaturated fats

❶ Place three drops of oil or melted fat in a test tube.
❷ Put the test tube in a 250 cm³ beaker of hot water.
❸ Add 3 cm³ of bromine water. *Wear eye protection.*
❹ Shake the tube gently to mix the contents.
❺ Time how long it takes the yellow/orange colour to fade away.
❻ If the colour fades away in a few seconds the fat or oil is 'high in unsaturates'.

Bromine reacts with unsaturated fat to form a colourless substance.

7.6

Why do we grow a lot of oilseed rape?

The name 'rape' is from the Old English word for turnip, *rapum*.

Oilseed rape is a very useful crop. It has yellow flowers and smells of honey. Half the weight of its seeds is oil. After storing the seeds, they are crushed between rollers to extract the oil. There is so much oil in the seeds that no further processing is needed.

The material left after removing the oil is rich in protein. It is good for cattle food.

Oilseed rape is grown throughout the world. It is used for animal feed, plant oil and **biodiesel**. Rape seed oil is turned into diesel fuel by chemically pulling the plant oil molecules apart. Fuel from crude oil is running out, so it very useful to find a way of making fuel from plant oil.

Figure 7.16 Yellow flowers of oilseed rape

Biodiesel is a motor fuel that has been made by chemically changing rapeseed oil.

Genetically modified crops have been changed by artificially moving genes from one organism to another.

22 Describe an oilseed rape plant. What would you see and smell?

23 How is the oil extracted from rape seeds?

24 a) What use is made of the material left after the oil is extracted?
   b) Why is this material so useful?

25 What non-food use is made of rape seed oil?

Activity – The GM controversy

A seed company has produced and sold **genetically modified** (GM) rape seeds. These plants are resistant to the effects of some weedkillers. This means that farmers can spray weedkiller on the fields to destroy the weeds. Their GM oilseed rape crop will not be damaged.

The company finds out that some farmers are growing their GM oilseed rape but they didn't buy the seeds. The farmers say that GM pollen was carried into their fields. It combined with the non-GM oilseed rape which they had planted.

Other farmers find unwanted GM plants in their fields that are not killed by weedkillers.

Weedkillers destroy wildflowers as well as weeds. This leaves many insects such as bees and butterflies unable to survive because they depend on wildflowers.

Work in a group.
1 Make a list of the advantages of producing GM rape seed.
2 Make a list of the disadvantages of producing GM rape seed.
3 On balance, do you think production of GM rape seed should be increased or reduced? Explain your decision.

7.7 Alternatives to petrol and diesel

Petrol and diesel are likely to run out in your lifetime. Just imagine what might happen when petrol starts to run out.

Locally produced motor fuel might become the new growth industry. Small-scale factories could spring up in every town and city to produce alternatives to petrol and diesel. In the following activity, you have to decide which fuel you will back.

Figure 7.17 H$_2$ Eyes Cool. H$_2$ Eyes is hi-tech. She sells solar cells that turn water into hydrogen fuel. She can convert cars to run on it. She's light and bright!

Figure 7.18 Betty Biodiesel. Betty makes diesel motor fuel out of recycled plant oil from local farmers. She's keen and green!

Activity – Fuel for the future!

Read the arguments put forward by each salesperson for their fuel. Then answer the questions.

H$_2$ Eyes Cool is selling hydrogen and fuel cells
- The raw materials for hydrogen are sunlight and water.
- Photocells (solar cells) split up water to produce hydrogen and oxygen. The hydrogen can be used as a fuel. The oxygen is released into the air.
- Although photocells are expensive, they can always produce hydrogen as long as there is sunlight.
- The cells will only produce hydrogen in daylight hours.
- Hydrogen fuel is clean – the waste materials from the process are oxygen and water.
- Hydrogen vehicles don't have an engine that burns fuel, so there are no noisy exhausts.
- Hydrogen cars have an electric motor. The hydrogen from the photocells powers a fuel cell to make electricity.
- Hydrogen can be produced from natural gas instead of using photocells if necessary.

Betty Biodiesel is selling biodiesel
- Diesel engines do not have to be modified to burn biodiesel.
- Biodiesel can be mixed with ordinary diesel to make it last longer.
- Biodiesel produces 80% less carbon dioxide than ordinary diesel.
- Biodiesel does not produce SO$_2$, so acid rain will be reduced.
- Unburnt hydrocarbons cause smog. Petrol and ordinary diesel leave ten times more unburnt hydrocarbons in the air than biodiesel.
- Biodiesel produces lubrication, so engines last longer.
- Biodiesel has been used in Europe for 20 years.
- Biodiesel is safe to handle and transport. It is not explosive like petrol.
- Biodiesel is made from farm crops such as soya, sunflowers and oilseed rape.

Figure 7.19 Sugar Dude. Sugar runs a fermenting plant. She makes ethanol fuel for motor vehicles. She's cooking up a clean future!

Sugar Dude is selling ethanol fuel

- Ethanol is made by fermentation – a natural process in rotting fruit.
- Crops such as maize, barley or potatoes can be used instead of fruit.
- Burning ethanol does not increase the amount of carbon dioxide in the atmosphere because the ethanol was produced from plants which had taken carbon dioxide from the air.
- Ethanol burns very cleanly and produces no polluting gases.
- Ethanol is not a hydrocarbon, so there are no unburnt hydrocarbons to cause smog.
- Ethanol mixes with water and does not pollute it.
- Ethanol is very flammable and bursts into flames easily.
- Pure carbon dioxide is produced by fermentation. This can be collected and used for fizzy drinks.
- After fermentation, the 'mash' (solid residue) can be made into animal feed.
- Ethanol keeps your engine clear and clean.

Work as a group.

1. Discuss each of the fuels and compare them with petrol and diesel. Think about the benefits and drawbacks of each new fuel. For example, consider cost, cleanliness, availability, supplies, whether cars would need to be changed.
2. Make a table listing the benefits and drawbacks of each fuel.
3. Ask someone in the group to make a short (3 min) presentation on behalf of each fuel.
4. Decide as a group which fuel you think is best for the future.

Summary

✓ Plants can provide oil from seeds and fruits.

✓ Mixtures of oil and water are called **emulsions**.

✓ Emulsions are widely used in cooking.

✓ Frying in oil cooks faster than boiling in water. It often makes the food more tasty.

✓ **Additives** are chemicals used to improve the storage, flavour and texture of food. But some additives have been linked with behaviour problems especially in children.

✓ Plant oils can be made into margarine, which is an alternative to butter.

✓ **Unsaturated fats** from plant oils are more healthy than **saturated fats** from animals. The most healthy diet is low in all fats.

✓ Plant oils can be made into a motor fuel called **biodiesel**.

1 Copy and complete the sentences below using the words in this box.

> grow   oil   starch   photosynthesis

A plant collects energy during _____. This makes _____ in the leaves of the plant. Plants also store energy as _____ in their seeds. This energy store is to help the seed _____.

(4 marks)

2 Here is a recipe for homemade mayonnaise.

Homemade mayonnaise

Ingredients
1 egg yolk
150 cm³ olive oil
10 cm³ grainy mustard
50 cm³ fresh lemon juice
salt and freshly ground black pepper to taste

Method

1 Place the egg yolk in a large bowl.
2 Gradually whisk the oil into the egg yolk.
3 When the egg and oil are well combined, whisk in the mustard and lemon juice.
4 Season the mayonnaise with salt and pepper.

a) What is the plant oil in the mayonnaise recipe? *(1 mark)*
b) What is the watery solution in the mayonnaise? *(1 mark)*
c) What are the emulsifiers in the mayonnaise? *(1 mark)*
d) How do the emulsifiers prevent the oil and water from separating? *(1 mark)*

3 Foods with additives look better, taste stronger and last longer. Who benefits most from this – the food producer, the shopkeeper or the customer who eats the food?

Copy and complete Table 7.2. Add as many points as you can in each box. Try to put something in each box. Then copy and complete the following sentence. *(6 marks)*

I think the person who benefits most is the _____. This is because _____. *(1 mark)*

5 Copy and complete the sentences below using words from this box.

> animal   kitchen   dressings   food
> pressing   biodiesel   oil   crude
> fruits   seeds

Many plants produce _____ or seeds with oil in them. The _____ is an energy store when the seed begins to grow. The plant oil can be extracted by _____ and heating the fruit or _____.

Plant oils have lots of uses. They are used in the _____ as a healthier alternative to _____ fat. They are used to make _____ more interesting and tasty. They are used in _____ such as mayonnaise.

More and more plant oil is used to make other things such as _____ for motor fuel. As _____ oil runs out, plant oils will be more important. *(10 marks)*

	Food producer	Shopkeeper	Person who eats the food
Advantages of food additives			
Disadvantages of food additives			

Table 7.2

Chapter 8
What changes have occurred in the Earth and its atmosphere?

At the end of this chapter you should:

✓ know that the Earth has a layered structure consisting of a crust, a mantle and a core;

✓ know that the Earth's crust is cracked into a number of very large tectonic plates;

✓ understand how the slow movement of these tectonic plates can cause earthquakes and volcanoes;

✓ understand why scientists cannot predict accurately when earthquakes and volcanic eruptions will occur;

✓ know the approximate percentages of gases in the Earth's atmosphere;

✓ be able to explain changes in the atmosphere since the Earth was formed;

✓ understand how and why the Earth's atmosphere is still changing.

Figure 8.1 This photo of the Earth was taken from space. It looks like a blue and white planet. Can you see continents, oceans and white clouds which form part of the atmosphere? In this chapter we will study the Earth and its atmosphere.

8.1

The Earth and its atmosphere

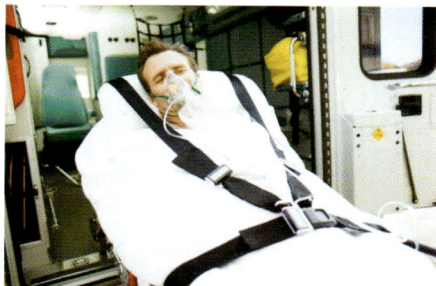

From the last three chapters, we know that the Earth and its atmosphere provide the raw materials for everything we need.
- The Earth itself provides crude oil, rocks for building materials and metal ores.
- The seas provide salt (sodium chloride) and fish to eat.
- The atmosphere provides important gases including oxygen and nitrogen.
- Living things (animals and plants) provide foods and raw materials for clothing and medicines.

People and animals need oxygen to breathe. Usually there is enough oxygen in the air around us. Sometimes people need extra oxygen to help them breathe. This pure oxygen is separated from other gases in the atmosphere by **fractional distillation** of liquid air.

Figure 8.2 Some people need extra supplies of oxygen to survive. In hospital, people who have difficulty in breathing and very small (premature) babies are given oxygen through a mask. Deep-sea divers and mountaineers carry extra oxygen in cylinders on their backs.

❶ Write the following statements (A, B, C and D) in the order they occur when oxygen is separated from argon and nitrogen in the atmosphere.
 A Allow liquid air to warm up.
 B Compress and cool air.
 C Nitrogen boils off first, then argon, then oxygen.
 D Oxygen liquefies first, then argon, then nitrogen.

8.2

Layers of the Earth

Fractional distillation is when a liquid is boiled and then condensed back to a liquid. The components of the liquid have different boiling points. They vaporise at different temperatures and so are separated out. They are cooled, condensed and collected. See Section 6.1.

The Earth is a huge ball of rock, iron and nickel. It has a radius of about 6400 km. The Earth formed about 4500 million years ago. At first, it was a mass of molten rock and metals. Over millions of years, the Earth cooled down. At the same time, heavier materials sank towards the centre of the Earth and less dense materials rose to the surface. This resulted in a layered structure (Figure 8.3).

There are three clear layers in the Earth:
- There is a thin outer **crust** of less dense material on the surface of the Earth. The crust is about 50 km thick.
- There is a **mantle** of moderately dense rock about 3000 km thick. The mantle extends about halfway to the centre of the Earth. Temperatures in the mantle range from 1500 °C to 4000 °C. The hottest rocks in the mantle form a very viscous liquid called **magma**. The magma changes shape and flows very slowly.
- There is a **core** of very dense iron and nickel with a radius of 3500 km (about half the Earth's radius).

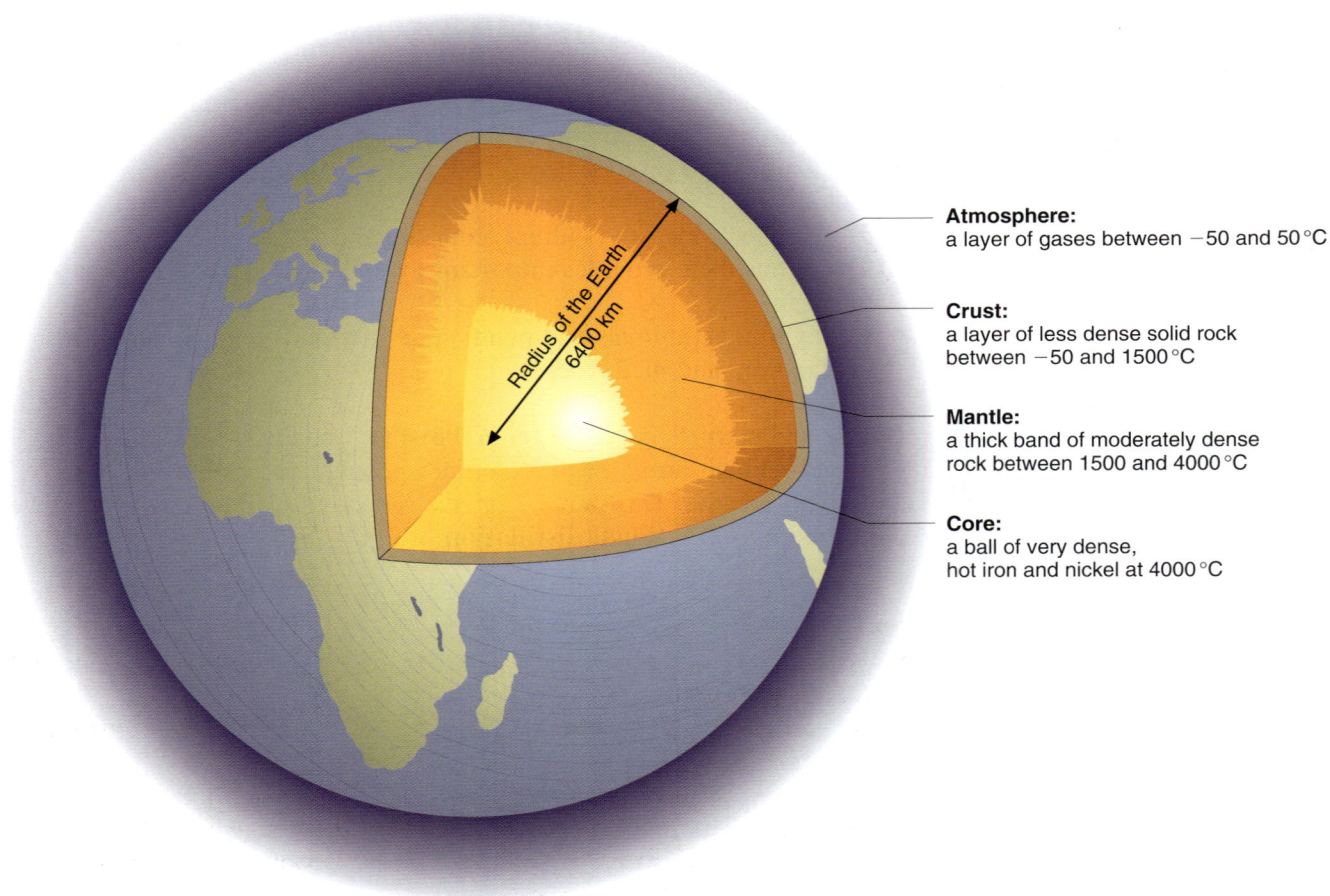

Radius of the Earth 6400 km

Atmosphere:
a layer of gases between -50 and $50\,^{\circ}\mathrm{C}$

Crust:
a layer of less dense solid rock between -50 and $1500\,^{\circ}\mathrm{C}$

Mantle:
a thick band of moderately dense rock between 1500 and $4000\,^{\circ}\mathrm{C}$

Core:
a ball of very dense, hot iron and nickel at $4000\,^{\circ}\mathrm{C}$

Figure 8.3 Layers of the Earth

❷ a) How do the temperatures of the layers in the Earth change towards the centre?

b) How do the densities of the layers in the Earth change towards the centre?

❸ a) Why did iron and nickel move into the core as layers formed in the Earth?

b) In the last 100 years, average temperatures on the surface of the Earth have risen slightly. Why do you think this has happened?

c) Do you think that temperatures on the Earth will continue to rise? Explain your answer.

Most of the evidence for these three layers in the Earth comes from the study of shock waves from earthquakes. Shock waves from earthquakes are reflected differently by the three layers. The reflections show the crust, the mantle and the core very clearly.

More detailed study shows that the core can be split into two – an **outer core**, which is liquid; and an **inner core**, which is solid.

Above the Earth there is a fourth layer – the **atmosphere** – which is a layer of gases about 100 km thick.

Why is the Earth's core so hot?

Figure 8.4 shows the main reason why temperatures remain high inside the Earth. Some rocks in the Earth, particularly granite, contain radioactive atoms of elements such as uranium and potassium (see Sections 11.8 and 11.9). As these radioactive atoms break up (decay), energy is released as heat and electromagnetic radiation. This energy helps to keep temperatures high inside the Earth.

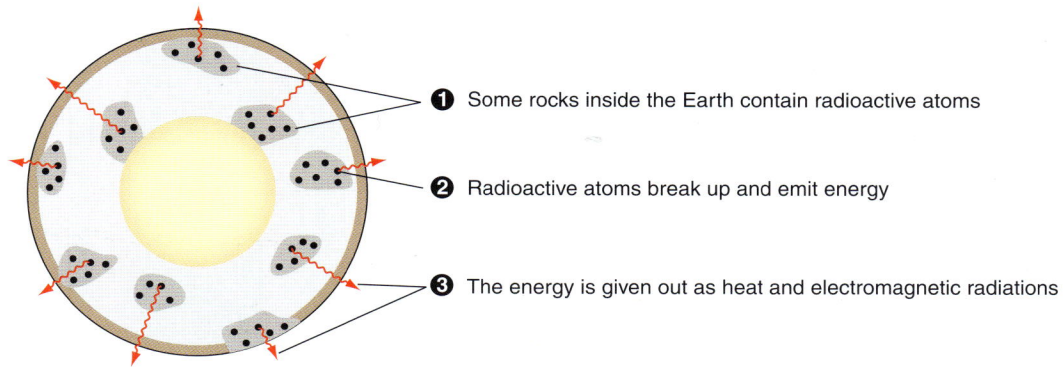

Figure 8.4 Explaining the high temperatures inside the Earth

① Some rocks inside the Earth contain radioactive atoms

② Radioactive atoms break up and emit energy

③ The energy is given out as heat and electromagnetic radiations

8.3 How is the Earth changing?

Before 1915, most scientists thought that:
- the Earth contracted and shrunk as it cooled down;
- the shrinking caused the Earth's crust to wrinkle (Figure 8.5);
- the shrinking and wrinkling of the Earth's crust caused mountains and valleys to form on the Earth's surface;
- the hard, solid rocks in the Earth's crust prevented any movement of the continents.

In 1915, the German weatherman Alfred Wegener suggested that the continents of the Earth were moving. At first, people laughed at his ideas because they seemed impossible. But, as people listened to Wegener's evidence, his ideas didn't seem quite so impossible after all.

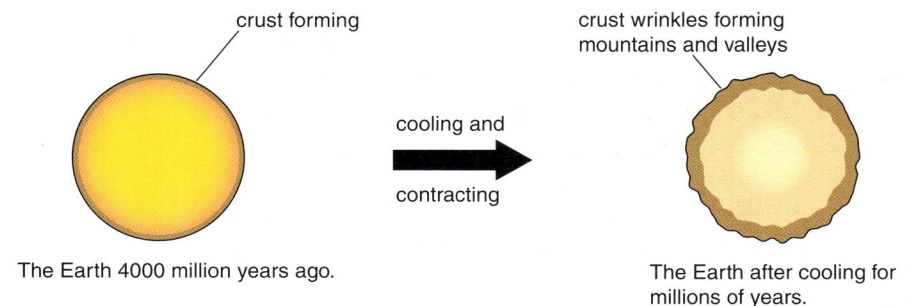

crust forming

crust wrinkles forming mountains and valleys

cooling and contracting

The Earth 4000 million years ago.

The Earth after cooling for millions of years.

Figure 8.5 Early ideas about how mountains and valleys formed on the Earth's surface

Activity – Wegener's evidence for moving continents

Continental fit
Wegener noticed that the continents looked like pieces of a giant jigsaw puzzle that could fit together. For example, it was easy to see how South America fitted into the coastline of West Africa (Figure 8.6).

① Use tracing paper to trace the outline of South America in Figure 8.6. Carefully cut out your outline. How well does it fit into the coastline of West Africa?

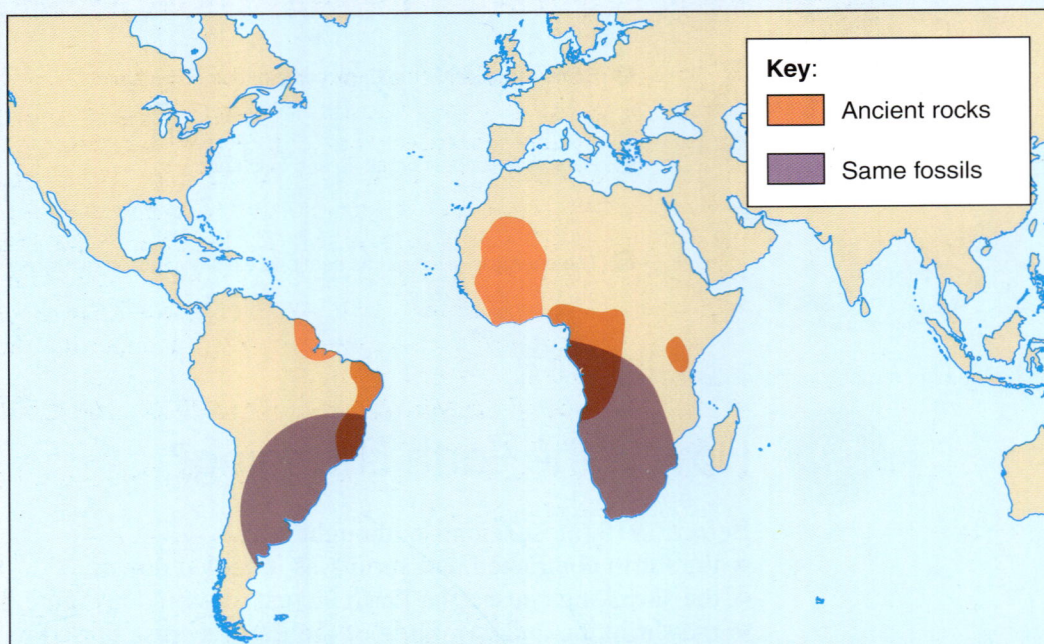

Figure 8.6 Wegener's evidence for the movement of South America and Africa

❷ Most geologists (people who study rocks) told Wegener that this was just a fluke. How could continents move sideways by thousands of kilometres? What do you think?

Mountain chains

Wegener also noticed that mountains with similar rocks were found in South America and Africa. These mountain chains would be close together if the continents formed one giant jigsaw (Figure 8.6).

❸ Using your outline of South America, shade the areas with ancient rocks as in Figure 8.6. Fit your outline as snugly as possible into the West Coast of Africa on Figure 8.6. How neatly do the areas with ancient rocks in South America and West Africa line up?

❹ The experts told Wegener that the line up of ancient rocks in South America and Africa was not exact. Is that right?

Fossil records

Animals and plants can only survive in areas where the climate and habitat suits them. Wegener noticed that fossils of the same animals and plants had been found in South America and Africa. But the climates and habitats of these two continents were very different. And these areas with the same

fossils were very close to each other if he fitted the continents together (Figure 8.6).

❺ Shade the area with the same fossils on your outline of South America. Now fit this outline again into the West Coast of Africa. How well do the areas with the same fossils in South America and West Africa line up?

❻ Those who disagreed with Wegener said there was probably once a land bridge connecting South America and West Africa. Is this possible?

From observations like those you have just made, Wegener suggested that:

- About 200 million years ago, all the continents were joined together as one super-continent. Wegener called this super-continent **Pangaea** from Greek words meaning 'all lands'.
- Then, over millions of years, Pangaea broke up and the continents slowly moved apart. Wegener called this movement **continental drift**.

❼ Wegener used fossils from South America and Africa to work out that the break-up of Pangaea happened 200 million years ago. Explain how you think he did this.

❽ Why do you think Wegener's theory of continental drift (crustal movement) was not accepted for many years?

8.4 Why are the continents moving apart?

Wegener just couldn't persuade people to believe him. This was mainly because he could not explain how or why the continents moved apart. But, in 1930, scientists realised that **convection currents** (see Section 9.2) in the mantle of the Earth might cause the continents to move. Clearer evidence for these convection currents and for continental drift was found in the 1950s. During the 1950s and 1960s, the floor of the Atlantic Ocean was surveyed in detail. This led to some important discoveries.

- Running down the centre of the Atlantic Ocean there is a mountain ridge with volcanoes. This mountain ridge is under the sea. It is called the Mid-Atlantic Ridge (Figure 8.7).
- On either side of the Mid-Atlantic Ridge, there is a similar pattern of humps and hollows.
- Rocks at the top of the ridge are almost new. The rocks further away from the ridge are older.

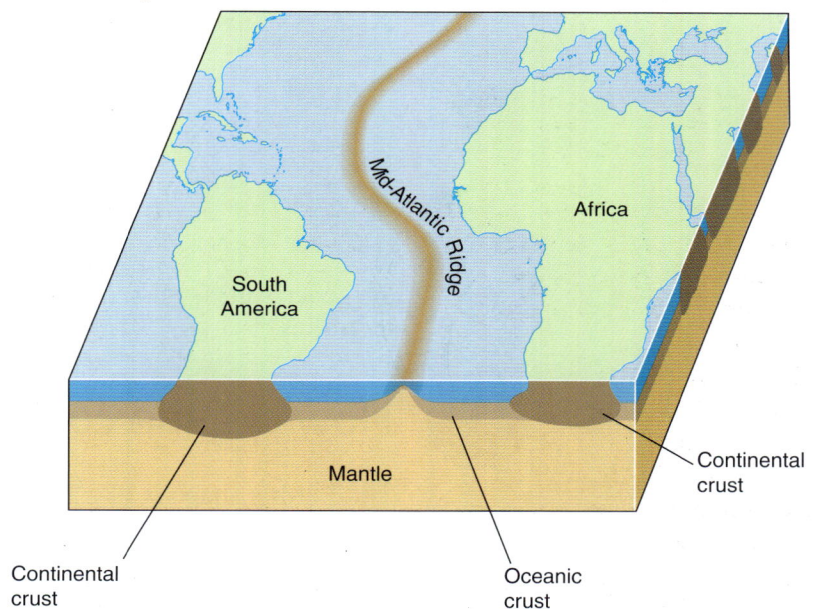

Figure 8.7 The Mid-Atlantic Ridge

These discoveries made scientists think that convection currents in the mantle of the Earth were causing lava to spew out from volcanoes onto each side of the ridge. As this happened over millions of years, the two halves of the sea bed moved apart.

The final proof of Wegener's ideas about moving continents came in the 1950s and 1960s.

During the 1950s, geologists studied the magnetism in rocks. As molten rocks become solid, the iron in them gets magnetised in line with the north–south direction of the Earth's magnetic field. To their surprise, the geologists found that every half-million years the Earth's magnetic field flips. The North Pole becomes a South Pole, and the South Pole becomes a North Pole.

Then, in 1963, surveys showed that the magnetism of rocks on one side of the Mid-Atlantic Ridge was an exact reflection of the other side (Figure 8.8). Rocks now hundreds of kilometres apart must have formed at the same time. They were formed on the ridge when magma spewed out of the volcanoes. Then they were pushed further and further apart over millions of years as new rocks formed at the ridge. And as the rocks moved apart, the continents of South America and Africa moved with them.

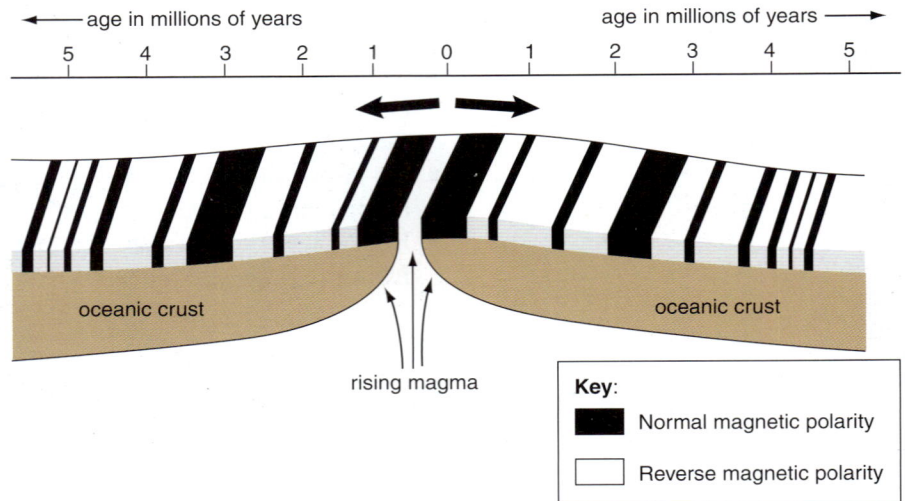

Figure 8.8 Magnetic patterns in the rocks on each side of the Mid-Atlantic Ridge

Wegener's ideas about moving continents resulted in the study of **plate tectonics**.

8.5 How are the continents moving?

The Earth's shape is like an orange – spherical like a ball, but slightly flattened at the poles. Its structure is like a badly cracked egg. The 'cracked' shell is like the Earth's thin crust. The 'egg white' is like the mantle and the 'yolk' is like the core.

The Earth's crust and the upper parts of the mantle are cracked and broken into a number of massive pieces called **tectonic plates**.

The Earth's core is very hot because of natural radioactive processes. This keeps temperatures in the mantle between 1500 °C and 4000 °C. These high temperatures cause convection currents in the liquid mantle. The currents cause the tectonic plates to move a few centimetres each year. The main tectonic plates and their direction of movement are shown in Figure 8.9 on a map of the world.

The movements of tectonic plates are normally very slow. But sometimes they can be sudden and disastrous.

4 The American Plate is moving away from the Eurasian and African Plates at a speed of 5 cm per year (Figure 8.9). Over millions of years, movements like this can cause massive changes on the Earth.
a) How many centimetres will the American Plate move away from the Eurasian and African Plates in one million years?
b) How many kilometres will it move away in one million years?
c) How much wider will the Atlantic Ocean be in one million years?

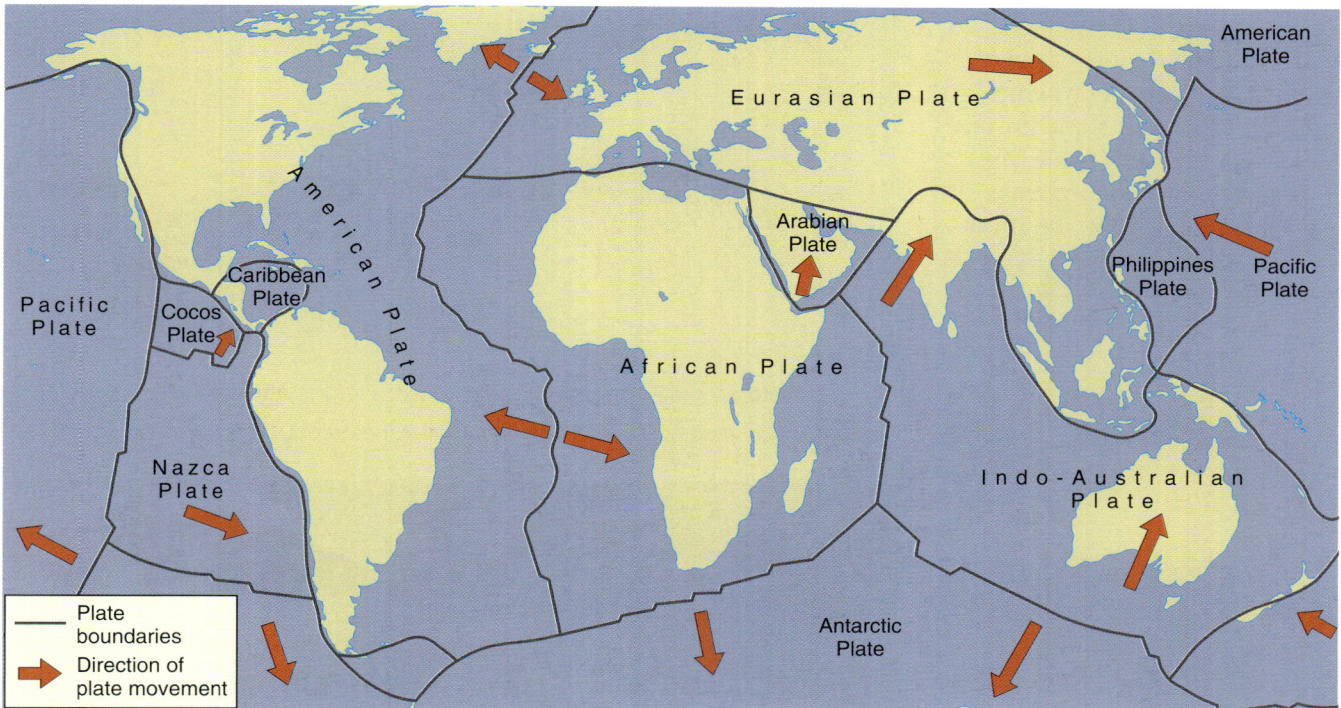

Figure 8.9 The main tectonic plates and their directions of movement

Plate tectonics explain many features on the Earth's surface. They explain why earthquakes and volcanoes happen. They even explain how mountains formed.

This is why studying plate tectonics is very important. Scientists have studied what happens when:
- plates slide past each other;
- plates move apart;
- plates move towards each other and collide.

What happens when plates slide past each other?

When two plates slide past each other, stresses build up in the Earth's crust. Massive forces are involved. The convection currents in the mantle are pushing tectonic plates one way or another. The forces are so great that they may cause the plates to bend. In some cases, the stresses build up and are then suddenly released. The Earth moves and the ground shakes violently in an **earthquake**. Sometimes, breaks appear in the ground (Figure 8.10).

These breaks in the ground when plates slide past each other are called **tear faults** (Figure 8.10b). The San Andreas Fault in California and the Great Glen Fault in Scotland are tear faults. The map of Scotland would look very different if the Great Glen Fault had not happened (Figure 8.11).

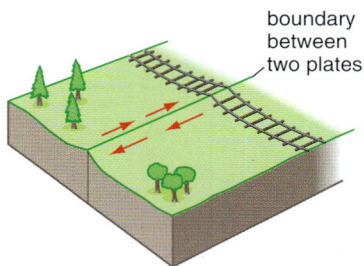

a) Plates in the Earth's crust are bent as they slide past each other.

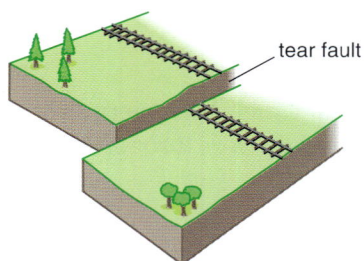

b) Stresses in the bent plates are suddenly released, the earth moves and the ground shakes violently in an earthquake. Breaks appear in the ground and a tear fault has formed.

Figure 8.10 An earthquake can happen when plates slide past each other.

❺ In California, most orange groves have trees growing in straight lines. In some groves, the trees are out of line, but they were not planted this way. Explain what you think happened.

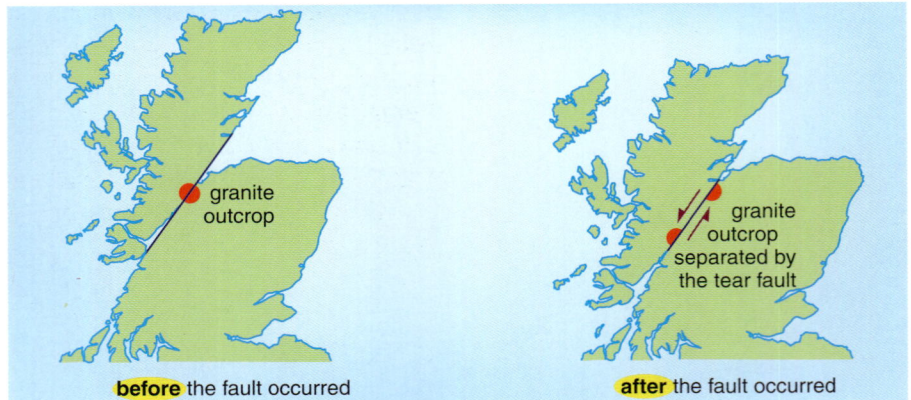

granite outcrop

before the fault occurred

granite outcrop separated by the tear fault

after the fault occurred

Figure 8.11 The map of Scotland before and after the Great Glen Fault

Figure 8.12 When an earthquake occurs below the sea bed, it can cause a giant wave called a tsunami. This is what happened in the Indian Ocean on 26 December 2004. More than 200 000 people were killed. Houses and other buildings were badly damaged.

Rift valley

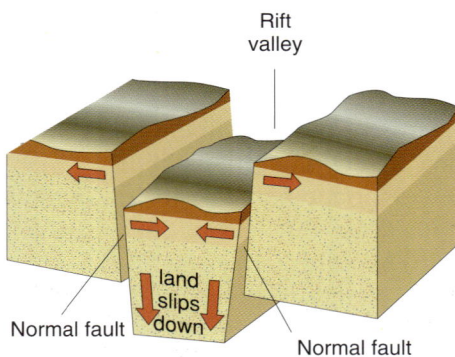

Normal fault

land slips down

Normal fault

Figure 8.13 As plates move apart, the land may sink into the crack. If there are two normal (vertical) faults near each other, a rift valley may form.

What happens when plates move apart?

When plates move apart, the crust is stretched. Cracks may appear in the Earth's surface. In some cases, hot molten magma rises through the cracks and erupts as a **volcano**.

If the plates move further apart, surface rocks sink and may get buried. When the land slips down it causes vertical faults called **normal faults**. When two vertical faults occur near each other, a rift valley forms (Figure 8.13).

Many volcanoes erupt without causing too much damage. But sometimes there is a super-volcano with terrible results. One of the most famous super-volcanoes is Krakatoa. It is on an island between Sumatra and Java in Indonesia. Krakatoa was dormant (not active) for centuries. Suddenly, it came 'back to life' in 1883. On 26 August 1883, two thirds of the island was blown up. Ash rose 27 km into the atmosphere. The explosions could be heard 4600 km away in Australia!

What happens when plates move towards each other and collide?

The Earth's crust under the continents is called **continental crust**. The crust under the oceans is called **oceanic crust**. The continental crust is usually thicker and contains less dense rocks, such as granite. The oceanic crust contains denser rocks, such as basalt.

When a continental plate and an oceanic plate collide, the denser oceanic plate sinks below the lighter continental plate. This process is called **subduction**. Parts of the oceanic crust may be forced into the mantle. Here, it melts to form magma. At the same time, the continental crust gets squashed and pushed into **folds** (Figure 8.14).

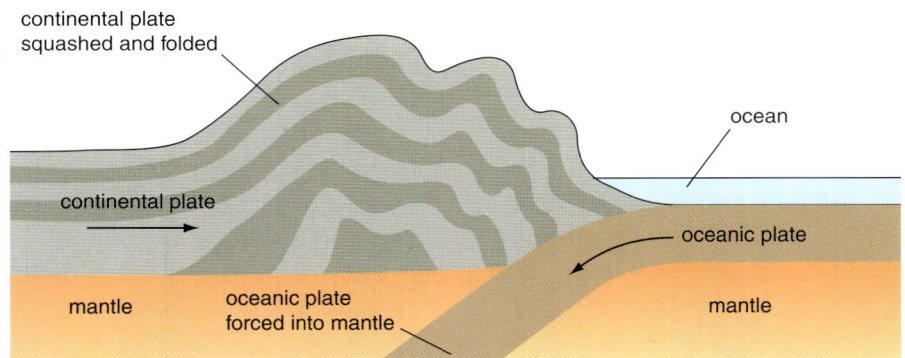

continental plate squashed and folded

ocean

continental plate

oceanic plate

mantle

oceanic plate forced into mantle

mantle

Figure 8.14 A continental plate and an oceanic plate moving towards each other

Over millions of years, this causes mountains to form. This is what happened and continues to happen in South America. The Nazca Plate is colliding with the American Plate forming the Andes Mountains.

If two continental plates or two oceanic plates move towards each other, they tend to collide head on. Layers in the plates become tilted and folded. This process has been happening for millions of years in the Alps, the Pyrenees and the Himalayas.

Sometimes cracks, rather than folds, appear when plates move towards each other. Land on one side of the crack is forced up above the other side which may then get buried (Figure 8.15). This is called a **reverse fault**.

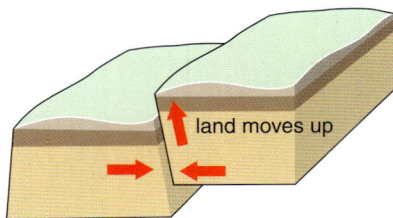

land moves up

Figure 8.15 When plates move towards each other, the Earth may crack forming a reverse fault.

6 Look at the photo of Stair Hole, Lulworth Cove, in Figure 8.16.

a) Draw a diagram of the land at Lulworth Cove. On your diagram, show the layers of sedimentary rock and the directions of the forces causing the fold. Label the fold.

b) Explain how the fold was formed.

Figure 8.16 Folds in the Earth's crust at Stair Hole, Lulworth Cove, in Dorset. These folds have been created by movement of the Eurasian Plate.

Activity – Earthquake

Breaking news – Massive Earthquake

We report the latest situation
BBC News 24
For further information go to:
www.geo.ed.ac.uk and search on 'quakes'
or www.eqnet.org

Use the internet websites in the newsflash above to find out about a recent large earthquake.

Now imagine you are the senior reporter for a news programme on TV.

Plan and write a report of about 250 words for the news programme.

Important things that you should mention are:
- the time and place of the earthquake;
- the number of people missing;

- the number of people killed, and those who are injured;
- the damage to homes and other buildings in the area;
- the size of the earthquake on the Richter scale.

Other things that you may wish to include are:
- whether there were any minor shocks after the main earthquake;
- whether the earthquake had been predicted;
- the dates of any previous earthquakes in the area;
- stories from people affected by the earthquake.

8.6 Can we predict when earthquakes and volcanic eruptions will occur?

Most earthquakes occur along or near the boundaries between tectonic plates. One of the areas most at risk from earthquakes is along the San Andreas Fault in California. Here, the Pacific Plate is sliding past the North American Plate.

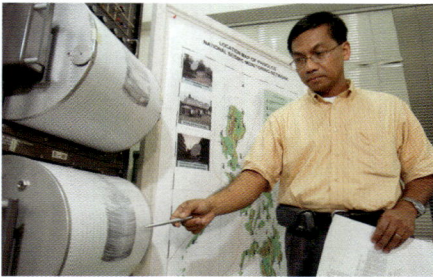

We can predict where earthquakes are likely to occur in the world. But we can't predict when they will occur. If we could, we would be able to move people out of their homes to safety and many lives could be saved.

Predicting earthquakes

Earthquakes occur suddenly and without warning. Predicting them is difficult. But there are signs to look for. And there are monitoring instruments that scientists can use.

- **Seismometers** can detect even the tiniest movements of the Earth. They are used in risky earthquake areas to pick up any minor shocks before the main earthquake.
- **The GPS (global positioning system)** locates positions on Earth to within a few centimetres. It is used to track the movement of plates.
- **Strainmeters** are used to measure the forces in rocks.
- **Animal behaviour** may change before an earthquake. Snakes come out of hibernation, rats leave their holes and cattle become restless.

Sometimes, scientists have been very successful in predicting earthquakes. For example, in 1975 Chinese scientists in Haicheng recorded increased earth movements on their seismometers. They also had reports that snakes were coming out of hibernation. People living in Haicheng were warned and told to leave their homes. At 7.30 p.m. an earthquake happened. Most of the buildings in Haicheng were destroyed, but thousands of lives were saved.

Predicting volcanic eruptions

It is much easier to predict when a volcano will erupt than when an earthquake will occur. Unlike earthquakes, volcanoes usually show warning signs before they erupt. Here are four warning signs.

- **Seismometers** detect small earth tremors caused by the movement of magma deep inside a volcano.
- **The GPS** shows changes in ground level around a volcano as pressure builds up inside it.
- **Air monitoring equipment** shows higher levels of sulfur dioxide near the mouth of a volcano. An increase in sulfur dioxide shows that the magma is rising.
- **Infrared cameras** on satellites detect an increase in ground temperature near a volcano. This may be due to magma rising towards the Earth's surface.

Figure 8.17 A seismograph used to monitor for the risk of earthquakes in the Philippines

Figure 8.18 The volcano on the island of Montserrat in the Caribbean had been quiet for 400 years. Then, in the summer of 1997, it suddenly erupted. The capital, Plymouth, had to be abandoned because of the flow of red hot lava.

These instruments give clues that a volcano might soon erupt. But predicting when the eruption will happen is more difficult.

In 1985, the ground rose by 2 m in only a few days near the volcano on Mount Vesuvius in Italy. About 40 000 people were moved out of their homes. Everyone waited for Mount Vesuvius to erupt. But it never happened.

> ❼ Explain why scientists cannot predict accurately when an earthquake will occur.
>
> ❽ Why are volcanic eruptions easier to predict than earthquakes?
>
> ❾ List three clues which can tell scientists that a volcano may soon erupt.

8.7

How has the Earth's atmosphere changed?

Scientists estimate that the oldest rocks on the Earth were formed about 4500 million (4.5 billion) years ago. This is usually taken to be the age of the Earth. There have been five key stages in the earth's atmosphere since its formation 4500 million years ago.

Stage 1 For 1000 million years after the Earth's formation, there was intense activity from volcanoes. Rocks decomposed, elements reacted and gases were given off. These gases formed the first atmosphere.

The early atmosphere was mainly carbon dioxide and water vapour. There were smaller proportions of methane (CH_4) and ammonia (NH_3) (Figure 8.19).

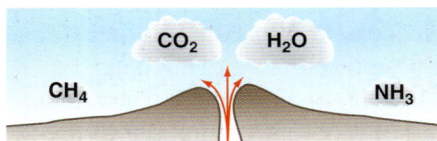

Figure 8.19 The early atmosphere on Earth was mainly carbon dioxide and water vapour.

Stage 2 As the volcanoes stopped erupting, molten rocks on the Earth's surface cooled. The temperature dropped still further and most of the water vapour condensed to form rivers, lakes and oceans (Figure 8.20).

Stage 3 Plants evolved and first appeared on the Earth about 3000 million years ago. As plants spread over the Earth, further changes took place in the atmosphere. Plants took in water from the earth and carbon dioxide from the atmosphere. They used these for photosynthesis and gave out oxygen (Figure 8.21).

Figure 8.20 As the Earth cooled down, most of the water vapour in the early atmosphere condensed to form rivers, lakes and oceans.

> ❿ There was no oxygen in the early atmosphere on Earth. What would have happened to any oxygen which did form? Remember that the Earth would be very hot.
>
> ⓫ On which planet in our Solar System today might you find conditions like those on the early Earth? Explain your answer.

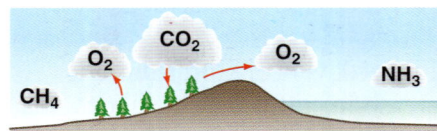

Figure 8.21 When plants appeared on the Earth, they took in carbon dioxide and water for photosynthesis and produced oxygen.

Stage 4 As oxygen collected in the atmosphere, flammable gases such as methane and ammonia reacted with it. These gases burned in the oxygen, producing more water, more carbon dioxide and nitrogen (Figure 8.22).

Stage 5 While oxygen was being added to the early atmosphere, carbon dioxide was being removed by photosynthesis in plants. It was also being removed by two other processes:

1   The formation of fossil fuels from the decaying remains of dead plants and sea creatures. The decaying remains contained carbon compounds. These carbon compounds were formed from carbon dioxide in the air by photosynthesis.

2   The formation of calcium carbonate as chalk and limestone in sedimentary rocks. The calcium carbonate came from the shells and bones of dead sea creatures. The food of these sea creatures contained calcium, carbon and oxygen compounds. The carbon and oxygen in these compounds was produced from carbon dioxide in the air by photosynthesis.

So, over billions of years, most of the carbon dioxide in the air gradually became locked up in fossil fuels and in sedimentary rocks as carbonates (Figure 8.23).

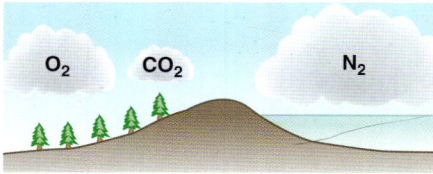

Figure 8.22 Methane and ammonia burned in the oxygen from photosynthesis. This produced more water, more carbon dioxide and nitrogen.

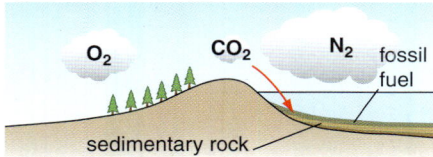

Figure 8.23 Over billions of years, carbon dioxide became locked up in fossil fuels and in sedimentary rocks as carbonates.

⑫ Write a word equation for the reaction of methane with oxygen in the Earth's early atmosphere.

⑬ Ammonia, in the early atmosphere, reacted with oxygen to form nitrogen and water vapour. Copy and complete the following equation for this reaction. Make sure that it balances.

$$4NH_3 + 3O_2 \rightarrow \underline{\quad}N_2 + \underline{\quad}H_2O$$

⑭ This question is about five gases – carbon dioxide, methane, nitrogen, oxygen and water vapour.
a) Which two of these gases formed most of the Earth's early atmosphere?
b) Which one of these gases will burn in oxygen?
c) Which two of these gases make up most of the atmosphere today?

⑮ How did oxygen get into the Earth's atmosphere?

8.8   # Our atmosphere today

Our atmosphere has been more or less the same for the last 200 million years. It is made up of:
● about four fifths (80%) nitrogen;
● about one fifth (20%) oxygen;
● small proportions of other gases including carbon dioxide, water vapour and noble gases.

Accurate percentages of these gases in dry air are shown in Table 8.1. The Earth is the only planet in the Solar System with oxygen in its atmosphere.

Gas	Percentage
Nitrogen	78.1
Oxygen	20.9
Argon	0.9
Carbon dioxide	less than 0.1
Neon	
Krypton	
Xenon	

Table 8.1 The percentages of gases in dry air

The noble gases

The noble gases are in Group O on the extreme right of the Periodic Table (Figure 8.24).

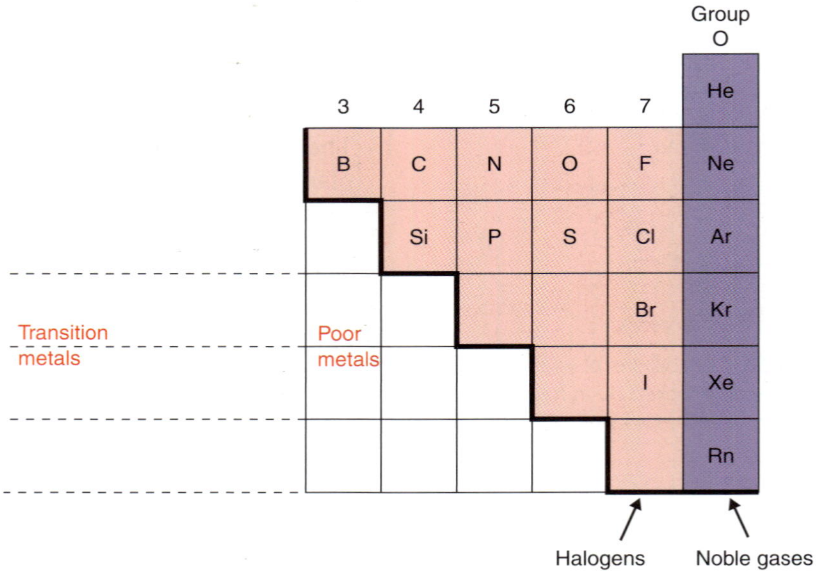

Figure 8.24 The position of the noble gases in the Periodic Table

The noble gases are all colourless and odourless. They also have very low melting points and boiling points. All the noble gases exist as separate single atoms. All the other gaseous elements (hydrogen H_2, oxygen O_2, nitrogen N_2 and the halogens) exist as molecules containing two atoms.

Until 1962, there were no known compounds of the noble gases. Chemists thought they were completely unreactive. Because of this, they were called the inert (unreactive) gases. Today, we know that they form several compounds. They are not inert, so we now call them the noble gases. The word 'noble' was chosen because unreactive metals like gold and silver are called noble metals.

Uses of the noble gases

Helium is used in balloons and airships because it has a low density and is non-flammable (Figure 8.25).

The noble gases produce a coloured glow when their atoms are bombarded by a stream of electrons. The stream of electrons can be produced with a high voltage across the terminals of a discharge tube, or from a laser. Neon and argon are used in discharge tubes to create advertising signs. Neon tubes give a red colour. Argon tubes give a blue colour (Figure 8.26).

Argon and krypton are used in electric filament lamps (light bulbs). If there is a vacuum inside the bulb, metal atoms evaporate from the tungsten filament when it gets very hot. To reduce this evaporation and make the filament last longer, the bulb is filled with argon or krypton. These gases are so unreactive that they do not react with the hot tungsten filament.

Figure 8.25 A technician releases a weather balloon filled with helium.

16 Hydrogen was once used for inflating balloons.
a) What was the special advantage of using hydrogen?
b) Why was the use of hydrogen dangerous?

17 a) Write down two reasons why argon and krypton are used in light bulbs (filament lamps).
b) Why is air unsuitable to use as the gas inside light bulbs?

Figure 8.26 *Many of the bright lights of Piccadilly are discharge tubes containing neon and argon.*

Activity – What changes are occurring in the Earth's atmosphere?

Table 8.2 shows the concentration of carbon dioxide in the atmosphere, the average worldwide temperature and the world population at intervals of 50 years since 1750.

	Year					
	1750	1800	1850	1900	1950	2000
Concentration of carbon dioxide in the atmosphere /percentage by volume	0.0278	0.0282	0.0288	0.0297	0.0310	0.0368
Average worldwide temperature/°C	13.3	13.4	13.4	13.6	13.8	14.4
World population/millions	350	500	1000	1500	3000	5500

Table 8.2

❶ Many people think that burning fossil fuels has increased the amount of carbon dioxide in the atmosphere.
a) Natural gas is the simplest fossil fuel. It is mainly methane with traces of ethane. Copy and complete the following equation for the burning of methane in oxygen.

$$CH_4 + \_O_2 \rightarrow CO_2 + \_\_H_2O$$

b) State two major uses of fossil fuels.
c) Name one important process which removes carbon dioxide from the atmosphere.

❷ Use the data in Table 8.2 to plot a graph which shows clearly that the level of carbon dioxide in the atmosphere is increasing.
a) In your graph, what is i) the independent variable, (ii) the dependent variable?

b) From your graph, describe the way in which carbon dioxide is increasing.

3 Many people believe that the increasing levels of carbon dioxide in the atmosphere are responsible for global warming.
   a) Give two reasons why we should worry about global warming.
   b) Which rows of data in Table 8.2 show that there might be a link between carbon dioxide and global warming?

4 Some people believe that global warming is largely the result of a world where there are more and more people. More people are using increasing amounts of fossil fuels. Do you think this is true? Say 'yes' or 'no' and explain your answer.

5 What should we in the UK be doing about global warming:
   a) at a personal or family level;
   b) at a national or government level?

Summary

✓ The Earth and its atmosphere provide the raw materials for everything we need.

✓ The Earth is nearly spherical with a layered structure. It has:
   ● a thin **crust**;
   ● a **mantle** extending almost halfway to the Earth's centre – it seems solid but it can flow very slowly;
   ● a central **core** about half the Earth's diameter, made of iron and nickel – the outer part of the core is liquid and the inner part is solid.

✓ The Earth's crust and the upper part of the mantle are cracked into a number of vast pieces called **tectonic plates**.

✓ The heat released by natural radioactive processes inside the Earth produces convection currents in the mantle. These convection currents cause the tectonic plates to move a few centimetres each year.

✓ The movement of tectonic plates is called **continental drift**. Normally it is very slow, but it can be sudden and disastrous. At the boundaries of the plates, it can result in earthquakes and volcanic eruptions.

✓ For the last 200 million years, proportions of the different gases in the Earth's atmosphere have stayed more or less the same:
   ● about four fifths (80%) nitrogen,
   ● about one fifth (20%) oxygen,

● small proportions of other gases including carbon dioxide, water vapour, argon and other noble gases.

✓ The **noble gases** such as helium and argon are in Group O of the Periodic Table. They are chemically unreactive. Helium is much less dense than air. It is used in balloons. Neon and argon are used in advertising signs. Argon and krypton are used in electric filament lamps (light bulbs).

✓ During the Earth's first billion years, there was intense volcanic activity.
   ● The volcanoes released gases which formed the early atmosphere. This was mainly carbon dioxide and water vapour. Smaller proportions of methane and ammonia were also present.
   ● As the Earth cooled down, the water vapour condensed to form rivers, lakes and oceans.
   ● When plants evolved, carbon dioxide and water were used up in photosynthesis, and oxygen was produced.
   ● Most of the carbon from the carbon dioxide in the air gradually became locked up in fossil fuels and in sedimentary rocks as carbonates.

✓ During the last 150 years, the level of carbon dioxide in the atmosphere has slowly increased. This is due to the burning of fossil fuels in industry (particularly power stations), in our homes and in our vehicles.

✓ The increase in the level of carbon dioxide in the atmosphere has resulted in **global warming**.

1 a) Figure 8.27 shows the layered structure of the Earth. Copy and complete Figure 8.27 by adding the three missing labels.

(3 marks)

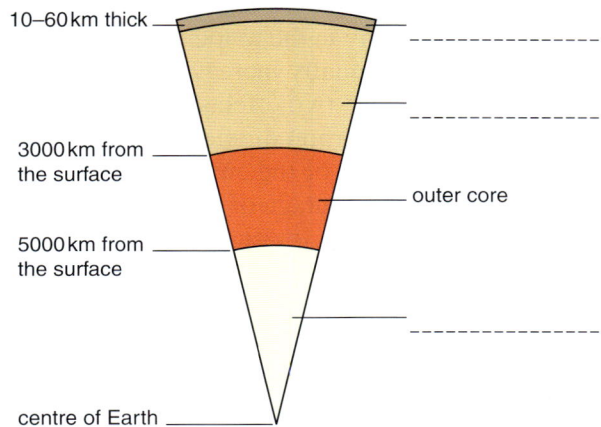

10–60 km thick

3000 km from the surface

outer core

5000 km from the surface

centre of Earth

Figure 8.27

EXAM QUESTIONS

b) The Earth's crust and upper mantle are cracked into a number of huge pieces called tectonic plates.
   i) Explain why these tectonic plates move.
   (2 marks)
   ii) Explain how the movement of tectonic plates can lead to earthquakes.
   (3 marks)

2 a) Figure 8.28 shows a cross-section through two tectonic plates A and B. Copy the diagram and add the following labels:

oceanic crust	continental crust	folds
mountains	sediments	mantle

(6 marks)

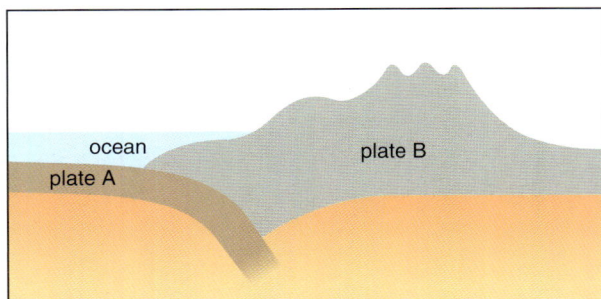

ocean

plate B

plate A

Figure 8.28

b) Put an arrow on your diagram to show the movement of plate B. *(1 mark)*
c) In one year, will plate A move a few millimetres, a few centimetres, or a few metres? *(1 mark)*
d) Why does plate A sink below plate B? *(1 mark)*
e) Mark a point X on your diagram where solid rock may be forming a thick, slow-moving liquid. *(1 mark)*

3 Match the terms A, B, C, D and E with the spaces 1–5 in the sentences below.
A continental drift
B convection currents
C land mass
D radioactive processes
E tectonic plates *(5 marks)*

Wegener suggested that, millions of years ago, there was a single, large __(1)__, which he called Pangaea. At some stage, Pangaea split up and the smaller pieces slowly moved apart. Wegener called this process __(2)__. Today, we call these slowly moving pieces of the Earth's crust __(3)__. They move because of __(4)__ in the Earth's mantle. The heat causing this movement comes from natural __(5)__.

4 The gases in the atmosphere have changed over billions of years since the first gases erupted from volcanoes.
a) Gases from volcanoes contain a lot of water vapour. Explain why our present atmosphere contains very little water vapour. *(4 marks)*
b) The Earth's early atmosphere contained a high proportion of carbon dioxide. Name two processes which reduced the amount of carbon dioxide in the atmosphere. *(2 marks)*
c) Oxygen did not appear in the Earth's atmosphere for about one billion (1 000 000 000) years. Why was this? *(2 marks)*

The Earth's early atmosphere

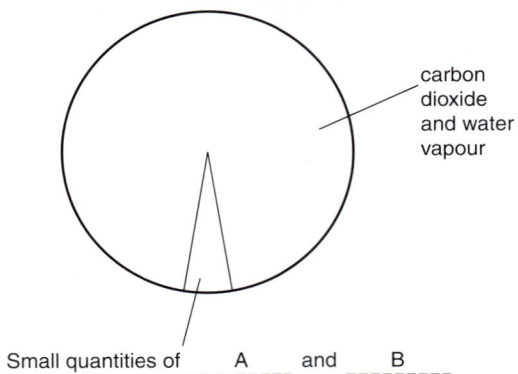

carbon dioxide and water vapour

Small quantities of ___A___ and ___B___

The Earth's present atmosphere

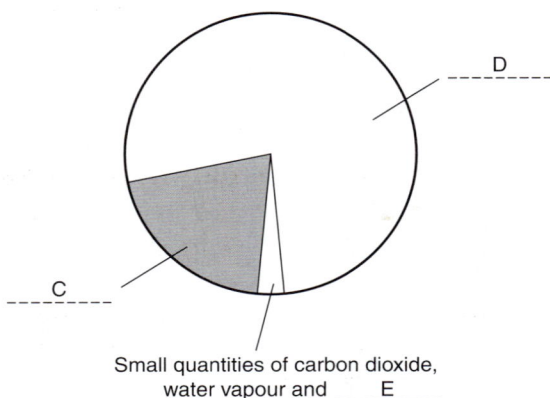

D

C

Small quantities of carbon dioxide, water vapour and ___E___

Figure 8.29

5 a) Name the gases labelled A to E in the pie charts in Figure 8.29. (*5 marks*)

b) In the last 150 years, the percentage of carbon dioxide in the atmosphere has increased from 0.03% to 0.04%.
   i) Why is the amount of carbon dioxide gradually increasing? (*2 marks*)
   ii) State two effects that increasing levels of carbon dioxide may have on the Earth or the environment. (*2 marks*)

c) The percentage of carbon dioxide in country air is often lower than that in city air. Why is this? (*1 mark*)

Chapter 9
How is heat transferred and what is meant by energy efficiency?

At the end of this chapter you should:

✓ understand how heat (thermal energy) is transferred;
✓ know the factors that affect the rate at which heat is transferred;
✓ know how to reduce the transfer of heat;

✓ understand how energy can be transformed (changed) from one form into another;
✓ know that an efficient device transfers more energy usefully than a less efficient device.

Figure 9.1 If we use energy efficiently we can save resources and save money.

How is energy transferred by radiation?

Radiation is the transfer (movement) of thermal energy (heat) by electromagnetic waves.

When you stand in the Sun, you feel warm. This is because you are absorbing (taking in) heat from the Sun. This heat comes from the Sun as **infra-red waves**. Infra-red waves are invisible. You can't see them but you can feel them. Anything that absorbs infra-red waves heats up. So infra-red waves are often called thermal radiation or just **radiation** for short.

Most of the space between the Sun and the Earth is completely empty – no air, nothing! We call an empty space like this a **vacuum**. Infra-red waves can travel through a vacuum. Light waves can too. Infra-red waves and light are examples of electromagnetic waves, which you will study in Chapter 11.

All bodies **emit** (give out) energy by thermal radiation. The hotter a body is, the more energy it emits by thermal radiation each second. Bodies also **absorb** (take in) thermal radiation.

Here the word **body** is used to mean an object (not a human body). A glass beaker, a metal spoon and a wooden toy are different objects, but they are all 'bodies'.

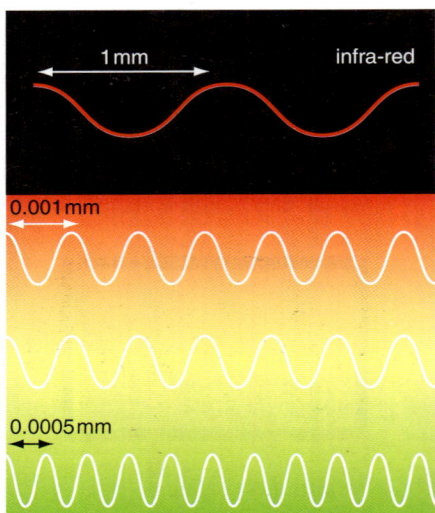

Figure 9.2 A very hot object emits infra-red waves and light waves. The light waves are visible as different colours and infra-red waves are invisible.

Emitting radiation

Dark-coloured matt surfaces emit radiation faster than light-coloured shiny surfaces (Figure 9.3). (A matt surface is dull and non-shiny.)

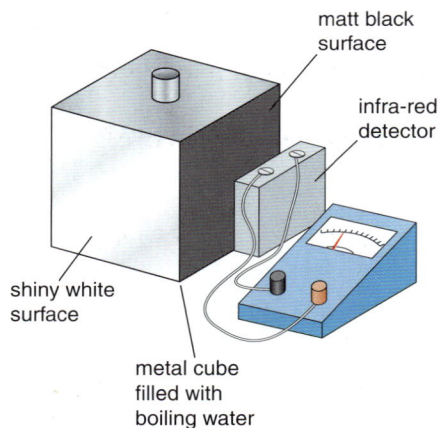

Figure 9.3 This metal cube contains very hot water. The water heats all the surfaces of the cube to the same temperature. The detector measures the amount of radiation emitted from each side of the cube. It gives the highest reading near the matt (non-shiny) black side of the cube. It gives the lowest reading near the shiny white side of the cube.

Variables

A variable is something that can vary, such as the different surfaces of the cube in Figure 9.3.

There are different kinds of variables:

Continuous variables can have any numerical value. The length of a piece of string, and the mass of a stone, are examples of continuous variables.

Ordered variables have a clear order of size or mass or length.

Small, medium and large pairs of socks are examples of ordered variables.

Categoric variables are different types of something. The different surfaces of the cube in Figure 9.3 are examples of categoric variables.

Later in this section, we will meet two other types of variable – independent variables and dependent variables (see also Section 5.7).

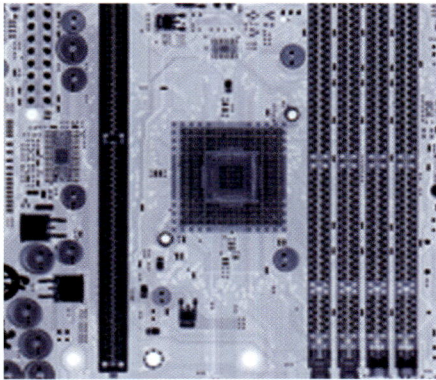

Figure 9.4 Electronic components are sometimes joined to a 'heat sink'. The matt black metal fins of the heat sink in the centre of this components board emit heat and keep the other components cool.

Absorbing radiation

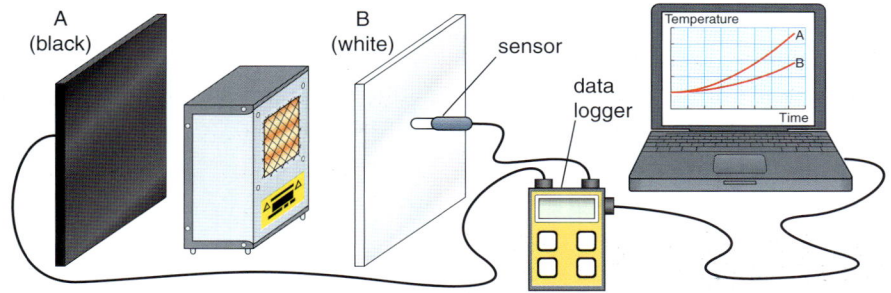

Figure 9.5 Two metal sheets (one black, the other white) are placed the same distance from an electric heater. Sensors linked to a computer measure the temperature of the sheets. The graph drawn by the computer shows that the temperature of the black sheet goes up faster.

Figure 9.6 The outside metal of a black car warms up faster in the Sun than the outside metal of a white car. A polished shiny car will not get as hot as an unpolished dull car.

Figures 9.5 and 9.6 show that:
- dark-coloured surfaces absorb thermal radiation faster than light-coloured surfaces;
- shiny surfaces reflect more thermal radiation than dull matt surfaces.

Good emitters of thermal radiation are also good absorbers of thermal radiation.

Figure 9.7 compares different surfaces as emitters, absorbers and reflectors of thermal radiation.

❶ What happens to the temperature of a hot potato when it emits more thermal radiation than it absorbs?

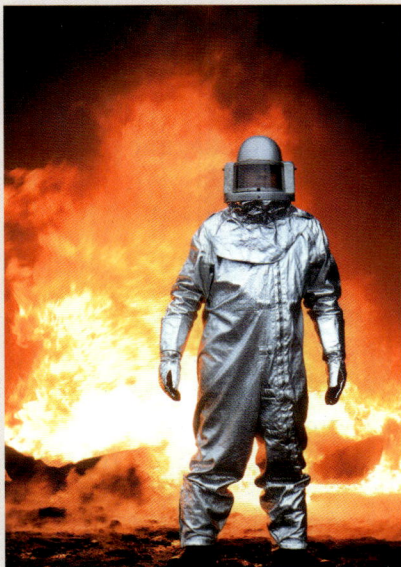

Figure 9.8 This fire-fighter is ready to tackle a blaze.

❷ Why is the suit worn by the fire-fighter in Figure 9.8 bright and shiny?

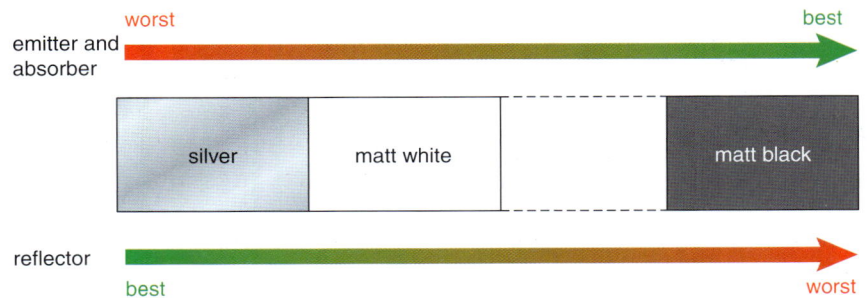

Figure 9.7 A summary of thermal radiation

❸ The window blinds in Ms Clymo's laboratory are matt black. The blinds are often closed. On a sunny day, Ms Clymo records the laboratory temperature. She does this every 15 minutes between 9.00 a.m. and 11.15 a.m. Her results are shown in Figure 9.9.

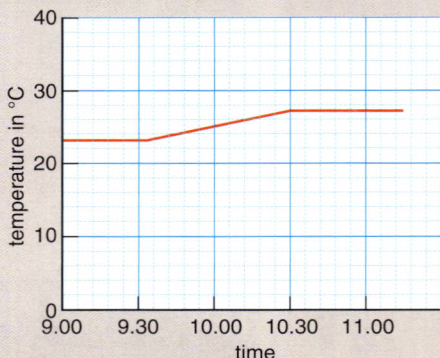

Figure 9.9 A graph showing the temperature inside Ms Clymo's laboratory

a) What time did Ms Clymo close the laboratory blinds? Give a reason for you answer.

b) Copy and complete the following sentence by choosing the correct phrase.
At 11.00 a.m. the temperature of the laboratory is *going up / staying constant / going down*.

c) At 11.00 a.m. the laboratory is absorbing 10 kJ of energy every second from the Sun. How fast is the laboratory losing energy?

d) Ms Clymo used a thermometer to measure the temperature. Why would it be better to use a temperature sensor and data logger?

e) It would be unwise to use Ms Clymo's results to predict the laboratory temperature at 3.00 p.m. The prediction would be unreliable. Explain why.

Activity – How does the temperature change as you move closer to a source of infra-red radiation?

To answer this question, Ted designed a simple experiment. The apparatus Ted used is shown in Figure 9.10.

Ted wrote this plan for his experiment.
- Put a thermometer with a blackened bulb 60 cm in front of the infra-red lamp.
- Measure the temperature.
- Switch on the lamp.
- Wait 2 minutes then take the temperature again.
- Move the thermometer 10 cm at a time towards the lamp.
- Each time the thermometer is moved, wait 2 minutes, then take the new temperature.
- Do the whole experiment again.

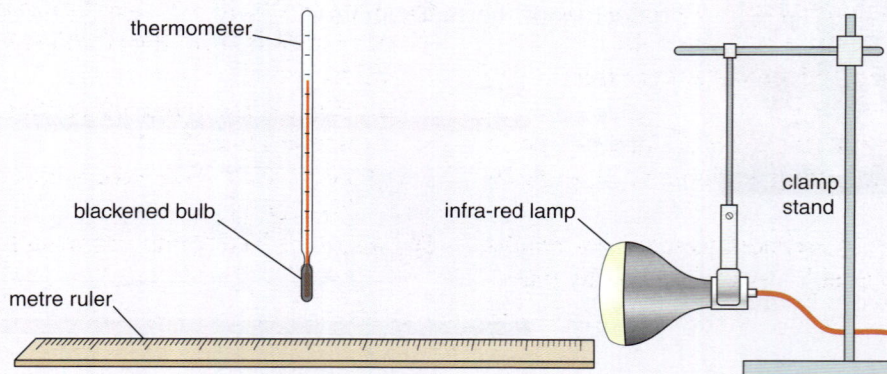

Figure 9.10 The apparatus used by Ted to investigate infra-red radiation

The results of Ted's experiment are shown in Table 9.1.

Distance from lamp in cm	Temperature in °C		
	First experiment	Second experiment	Mean
60	37	38	37.5
50	41	40	40.5
40	44	44	44.0
30	51	50	50.5
20	73	73	73.0

Table 9.1 The results of Ted's experiment

1. Why was it a good idea to use a thermometer with a black bulb rather than a clear bulb?
2. Why was it a good idea to wait 2 minutes before taking the temperature? Choose **one** of the following reasons.
   a) To give Ted time to write the results down.
   b) To give the thermometer time to reach a steady reading.
   c) To check that the temperature did not go down.
3. Give a reason why Ted got different results the second time he did the experiment.
4. Each time Ted moved the thermometer by 10 cm. This is called the **interval**. Why did Ted use an interval of 10 cm and not a bigger interval like 20 cm?
5. In an experiment, the **independent variable** is what you change or select. The **dependent variable** is what you measure when the independent variable changes.
   a) What is the independent variable in this experiment: distance or temperature?
   b) What is the dependent variable in this experiment?
   c) Is temperature in this experiment a categoric variable, a continuous variable or an ordered variable?
6. Ted repeated the experiment to make his results more **reliable**. Reliable results are results that

you can trust and that can be repeated. Are Ted's results reliable? Give a reason for your answer.

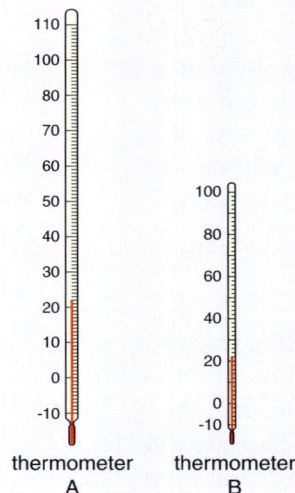

Figure 9.11 Thermometer A measures to the nearest 1 °C. Thermometer B measures to the nearest 2 °C. Thermometer A is more sensitive than thermometer B.

7. The **sensitivity** of an instrument is the smallest change that it can measure.
   Which one of the following describes the sensitivity of Ted's thermometer?
   • It measured to the nearest 0.1 °C.
   • It measured to the nearest 0.5 °C.
   • It measured to the nearest 1 °C.
8. Drawing a graph or putting results in a table often helps us to identify anomalous results. An anomalous result is one that does not fit the expected pattern. Are any of Ted's results anomalous?
9. A set of results has a certain range. The **range of results** is from the smallest to the largest value.
   a) What is the range of Ted's results for temperature (in °C)?
   b) How could Ted have increased the range of his results?
10. What has Ted found out from his experiment?

9.2

How is energy transferred by conduction and convection?

Conduction

Conduction is the way that thermal energy (heat) is transferred between materials that are touching. It is also the way that heat is transferred between different parts of the same substance, without the materials or the substance moving.

Do you sometimes walk around your home with bare feet? If you do, you will know that some floor surfaces feel much colder to your feet than others. For example, ceramic tiles feel much colder than carpet. Your feet feel the cold because they are losing energy in the form of heat to the tiles. Heat is transferred from your feet to the tiles by **conduction**.

All metals are very good conductors. But materials like plastic, wood and glass are poor conductors. Air and other gases are very poor conductors. Materials that are poor conductors are usually called **insulators**.

When one end of a metal bar is placed in a Bunsen flame, thermal energy (heat) from the flame is quickly transferred from the hot end to the cold end. This happens because metals contain tiny, vibrating particles (atoms) and electrons that are free to move (mobile electrons).

④ In winter, Daisy rides her bicycle to school. Later in the day she notices that the rubber handlebar grips feel warm but the metal frame feels cold. Explain why.

At the hot end of the metal bar (Figure 9.13a), energy from the flame makes the mobile electrons move faster. These electrons bump into other electrons making them move faster. In this way energy is conducted rapidly through the metal.

Energy is conducted along a metal bar by the movement of mobile electrons.

Figure 9.12 The handlebar grip feels warm but the frame feels cold.

a) In a metal energy is conducted quickly by fast-moving, mobile electrons.

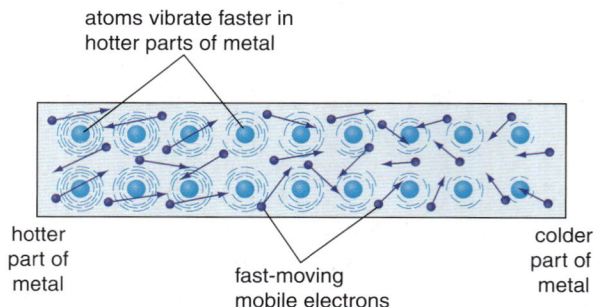

atoms vibrate faster in hotter parts of metal

hotter part of metal

fast-moving mobile electrons

colder part of metal

b) In an insulator, there are no mobile electrons, so energy is conducted slowly by the movement of electrons.

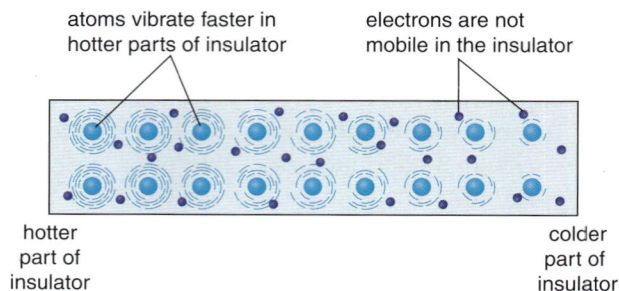

atoms vibrate faster in hotter parts of insulator

electrons are not mobile in the insulator

hotter part of insulator

colder part of insulator

Figure 9.13 Conduction in metals and insulators

> **Convection** is the transfer of thermal energy (heat) by the movement of a liquid or gas.

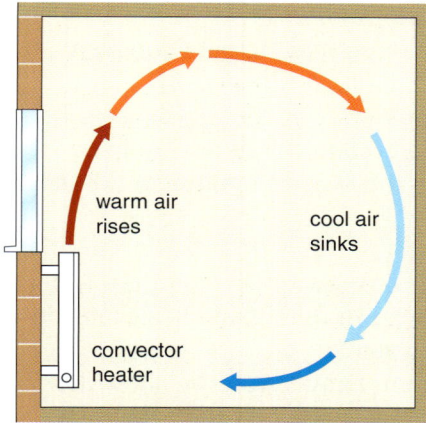

Figure 9.14 A heater creates an air flow in a room.

Figure 9.15 Trapped air keeps the rabbit warm.

Insulators such as plastics, wood and air have no mobile electrons. So the only way they can conduct is by the much reduced movement of electrons (Figure 9.13b). This is a slower process.

Convection

Convection is another way in which thermal energy (heat) can be transferred from one place to another. Convection happens in liquids and gases. During convection, the warmer liquid or gas tends to rise, transferring its extra energy with it.

Figure 9.14 shows how air can transfer energy by convection.
- Air closest to the heater warms up.
- This air expands as it warms up.
- The warm air is less dense than the colder air so it begins to rise.
- Colder, more dense air falls pushing warm air out of the way.

These movements of warm and cold air are called **convection currents**. Convection currents stop when all the air in the room is at the same temperature. Convection happens in a liquid in the same way as a gas.

Slowing down the energy transfer

Coping with the cold
Humans and other animals cope with the cold in various ways.

A seal has a thick layer of fat all around its body. Fat is a good insulator. The fat reduces the amount of heat lost each second from the seal's body.

Animal fur traps air and air is a very good insulator. When it is cold, animals will fluff up their fur to trap more air. In this way they can stay warm even in very cold weather.

Humans wear clothes to keep warm. When it's cold we put on more layers of clothes. The air trapped between each layer and in the fabric keeps us warm.

> **5** Copy and complete the sentences below using the words in this box.
>
> | conduction | convection | radiation |
> | --- | --- | --- |
>
> a) Heat transfer from atom to atom in a solid is called _____.
> b) Heat transfer by electromagnetic waves is called _____.
> c) Heat transfer by the movement of a gas is called _____.

6 Barbara runs a fast food shop called HotSnak. Some of her customers complained that her hot drinks go cold too quickly. Barbara decided to do a simple experiment to find a throw-away mug that keeps drinks hot for at least 10 minutes. Barbara tried four mugs, made from two different materials. She filled each mug with the same volume of boiling water. She put lids on two of the mugs. She waited 10 minutes, then took the new water temperature.

Table 9.2 shows the results of Barbara's experiment.
a) There were six variables in Barbara's experiment:
   Starting temperature and final temperature of the water; time; volume of water; type of material; lid or no lid.
   i) Barbara's experiment was a **fair test**. What variables did Barbara keep the same in order to make it a fair test?
   ii) The material of the mug is a **categoric variable**. This is because each mug belongs to one category or group. It was either made from cardboard or polystyrene.
   There was one other **categoric variable** in the experiment. What was it?

Mug	Material	Lid	Temperature after 10 minutes
A	Cardboard	Yes	60 °C
B	Cardboard	No	52 °C
C	Polystyrene	Yes	75 °C
D	Polystyrene	No	66 °C

Table 9.2

b) Copy and complete the following sentence by choosing the correct final word.
   If Barbara had tested each mug more than once, her results would have been more *accurate / precise / reliable*?
c) Which **two** results should Barbara compare before deciding to use cardboard or polystyrene mugs?
d) Which **two** results should Barbara compare before deciding if it is worth putting a lid on the mug?

7 Figure 9.16 shows a cut-away view of a hot water tank made of copper. An immersion heater inside the tank heats the water.
a) Copy the diagram. Draw arrows to show the convection currents caused by the heater.
b) Should the hot water tap be joined to pipe X or pipe Y? Explain your answer.
c) Copy and complete the sentences below.

Heat from the hot water passes through the walls of the copper tank by _____. The heat makes the _____ in the copper vibrate faster. The heat also makes the mobile _____ in the copper move _____. The expanded foam is a good _____. This is because the foam traps small pockets of _____.

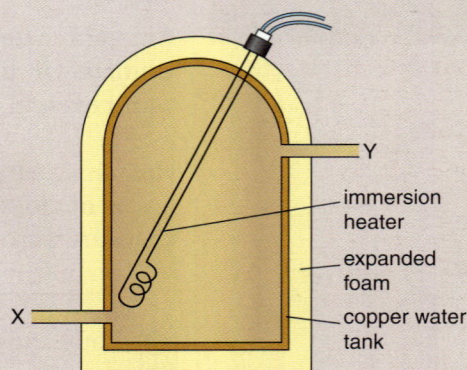

Figure 9.16

A **fair test** is one in which all the variables that are not being investigated are kept the same.

How does a vacuum flask keep hot drinks hot or cold drinks cold?
A vacuum flask can keep a hot drink hot or a cold drink cold for several hours. It does this by slowing down the energy transferred by conduction, convection and radiation.

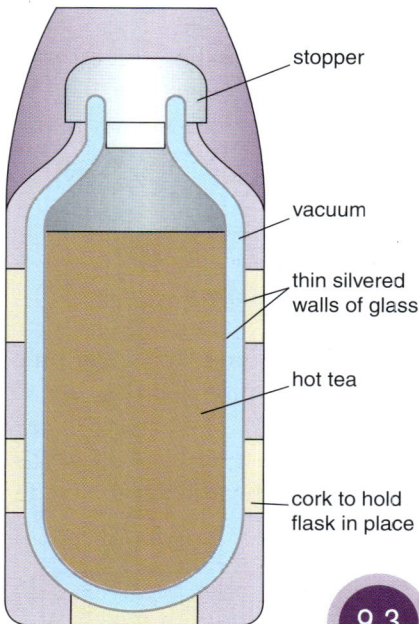

Figure 9.17 A vacuum flask keeps hot drinks hot and cold drinks cold.

An object loses heat when its temperature is higher than the materials around it.

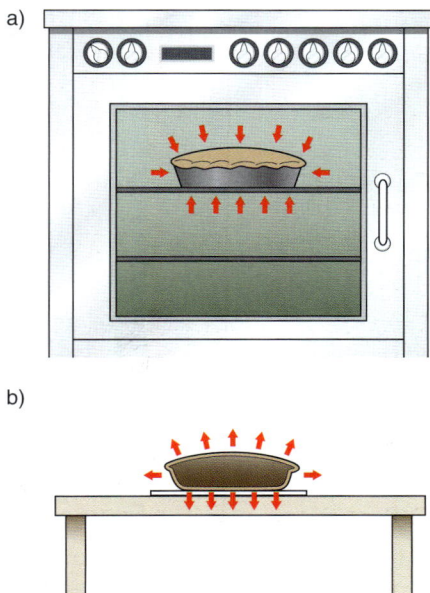

Figure 9.18 a) Heat is transferred to the pie from the hot air inside the oven.
b) Heat is transferred from the hot pie to the surroundings.

A vacuum flask slows down energy transfer by having:
- a vacuum between the double walls of the container – this reduces conduction and convection;
- walls with shiny surfaces – this reduces radiation;
- a stopper made of a good insulator like cork or plastic – this reduces conduction.

8 Some vacuum flasks are made from glass. Others are made from stainless steel. Which statement, A, B or C, gives a reason why vacuum flasks are made from glass rather than stainless steel?
A Stainless steel will not break as easily as glass.
B Stainless steel is a better heat conductor than glass.
C Stainless steel is shiny.

Do all objects lose heat at the same rate?

How fast an object loses heat depends on several things:
- the material the object is made of;
- the shape of the object;
- the size of the object;
- the difference in temperature between the object and its surroundings.

Type of material
Two hot objects made from different materials will cool down at different rates. If they start at the same temperature, they will have different temperatures after a few minutes. One object loses heat faster than the other.

Shape and surface area
The rate at which an object loses heat depends on its surface area. If the shape of the object is changed, its surface area will change and this will affect the rate of heat loss.

A motorbike engine would overheat if it did not lose heat fast enough. The cooling fins on the outside of the engine increase the rate at which heat is lost to the air. They do this by greatly increasing the surface area of the engine. This allows more cold air to come into contact with more hot metal.

Size

Objects of different size lose heat at different rates. Smaller objects cool down faster than larger objects with the same overall shape. For example, babies and small children start to feel cold much sooner than larger, heavier adults (Figure 9.19).

Temperature of the surroundings

The bigger the temperature difference between a body and its surroundings the faster the body warms up or cools down.

The graph in Figure 9.20 shows how the heat lost from a hot water pipe increases as the temperature difference between the pipe and the surrounding air increases.

When an object is touching a conductor it will lose heat faster than when it is touching an insulator.

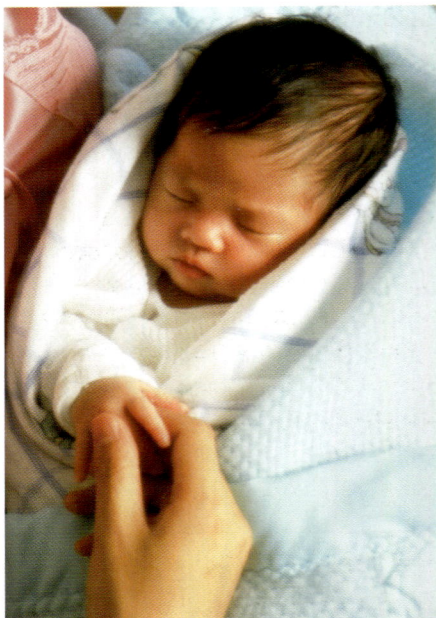

Figure 9.19 A baby feels the cold more quickly than an adult.

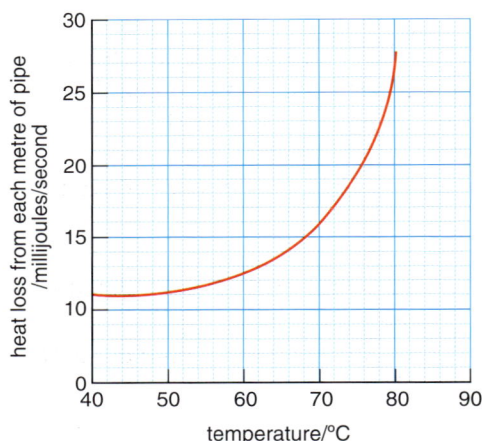

Figure 9.20 Heat loss from a hot water pipe

9 Figure 9.21 shows three ball-bearings in a beaker of hot water. When the ball-bearings are at the same temperature as the water they are taken out and put onto a table.

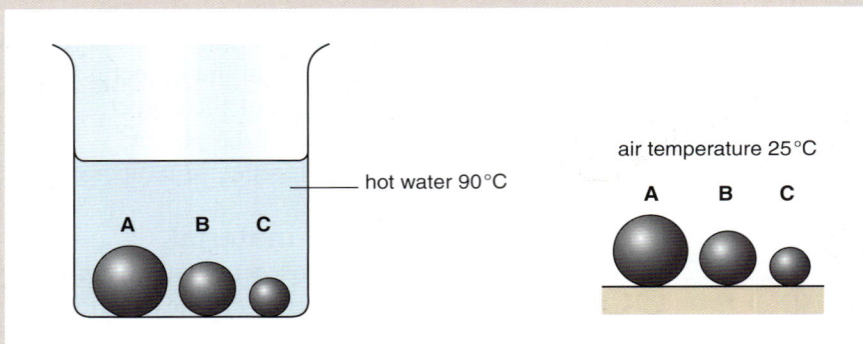

Figure 9.21

a) Which ball-bearing cools down the slowest? Explain your choice.

b) Copy and complete the following sentence by choosing the correct line in the box.
If the ball-bearings had been put in a refrigerator instead of on the table, they would have cooled down

> faster.
> at the same rate.
> slower.

c) Are the sizes of the three ball-bearings a continuous variable or an ordered variable?

9.4 How can the rate of heat loss be reduced?

Building services engineer

A building services engineer designs and installs the heating, lighting, electricity and ventilation systems for buildings such as homes, schools, factories and offices. A building services engineer makes sure that the buildings are safe, healthy and comfortable. Building services engineers try to make buildings as energy efficient as possible. They try to reduce the amount of energy wasted. This helps to cut down our use of energy resources. It also reduces any pollution the building may cause to the environment.

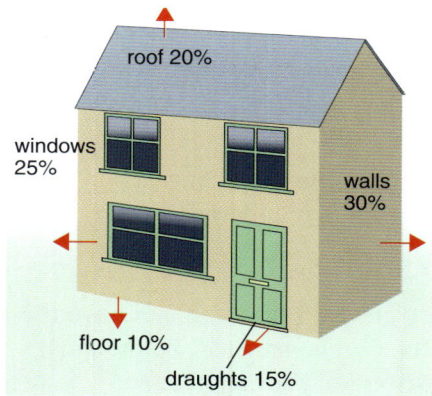

Keeping your home warm

Keeping your home warm is not just about turning on the heating. It's also about reducing the amount of heat lost from inside your home to the air outside. Building services engineers need to understand how energy (heat) is transferred. Then they can give advice to builders and householders on the best ways to make a house energy efficient.

Different amounts of heat are lost through the roof, the walls, the floor, the windows and the doors of a house. The percentages of the heat lost through these parts of an average house are shown in Figure 9.22.

Figure 9.23 shows some of the ways we can reduce the amount of heat lost from our homes.

Most ways of reducing heat loss work by trapping air. If air is trapped in small pockets it cannot move far. This means that heat loss by convection is greatly reduced. Air is also a good insulator so the amount of heat lost by conduction is reduced as well.

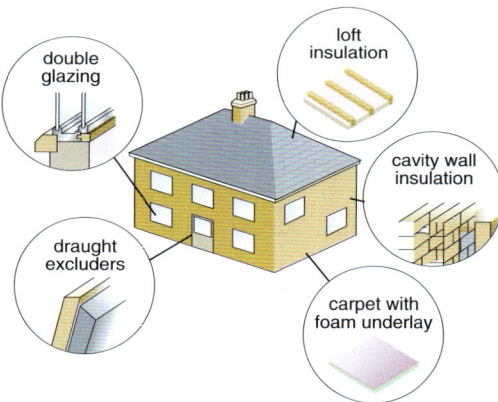

Figure 9.22 Heat loss from an average house

Figure 9.23 Reducing the heat lost from our homes

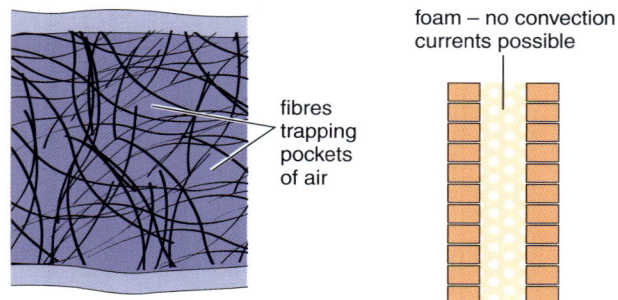

Figure 9.24 Air trapped by fibre wool and by foam

Double-glazed windows have two layers of glass with a thin layer of air in between. Both the air and the glass are good insulators. So the amount of heat lost by conduction is very small.

Draught excluders trap warm air inside a house and stop cold air getting in. But you need to be careful. Stopping all draughts is not good for you. It is very important that the air in a room changes at least once every hour. This replaces the oxygen that has been used up.

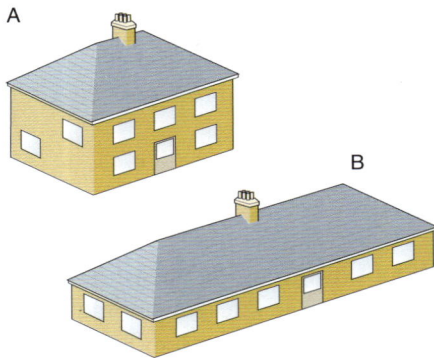

A

B

Figure 9.25

⑩ Copy and complete the following sentence.

You can reduce the heat lost from your home by having _____-glazed windows, laying carpet with a _____ underlay, fitting draught _____ and putting in loft _____.

⑪ Figure 9.25 shows two houses. They are the same size and have the same type of insulation. Which of the two houses will be the most expensive to keep warm? Give a reason for your choice.

⑫ Explain why fish and chips wrapped in layers of paper stay hot for some time.

⑬ Figure 9.26 shows a pudding called 'baked Alaska'. This pudding has ice-cream inside a crispy meringue topping. The pudding is baked in a very hot oven for a few minutes. Explain why the ice-cream does not melt.

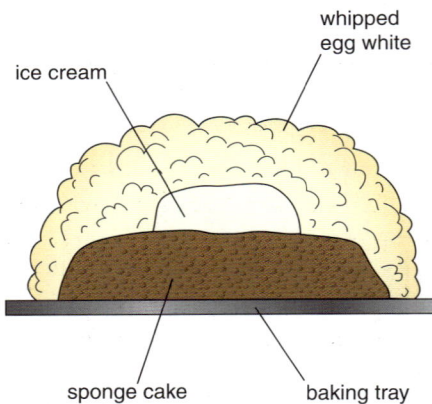

whipped egg white

ice cream

sponge cake

baking tray

Figure 9.26 A baked Alaska

Saving energy saves money

Improving the insulation in your home costs money. However, good insulation reduces the amount of energy needed to keep your home warm. This means that the heating bills are less, so you save money. But it can take a long time for the money saved on heating bills to pay for the cost of the insulation.

The time it takes to get back the cost of the insulation from the money saved on the heating bills is called the 'pay-back time'.

Example
Gurpal lives in a house that has no loft insulation. He wants to cut his heating bills, so he has just put in loft insulation. It cost £180. The insulation will save Gurpal £45 each year on his heating bill. What is the 'pay-back time' for the insulation?

$$\text{pay-back time} = \frac{\text{cost of insulation}}{\text{money saved each year}} = \frac{180}{45} = 4 \text{ years}$$

Gurpal will get back the cost of putting in the loft insulation in four years.

Figure 9.27 Thick curtains stop cold air getting into a room. They also trap air between the window and the curtains so less heat is lost by conduction.

The shorter the pay-back time and the longer the insulation lasts, the better it is for Gurpal. He saves more money, and the insulation is more cost-effective. Over time, the insulation will save far more money on energy bills than it cost to put in.

How cost-effective is Gurpal's loft insulation? If the insulation lasts 30 years and energy costs stay the same, the total amount saved on energy bills is £45 × 30 = £1350. So, after spending just £180, Gurpal will save £1350. That's cost-effective!

Closing the curtains is another way to reduce heat loss from your house (Figure 9.27). This costs nothing and it saves energy. Nothing could be more cost-effective than that!

❶❹ Table 9.3 gives the costs and savings for different methods of reducing heat loss from a home.

Work out the 'pay-back time' for each method of reducing heat loss.

Method of reducing heat loss	Cost to install	Savings made to yearly heating bill
Draught proofing	£42	£14
Hot water jacket	£16	£8
Double glazing	£4500	£150

Table 9.3

❶❺ Figure 9.28 shows two hot water tanks, A and B. Only tank B has an insulating jacket. The water in each tank is heated to 70 °C. The heaters are then switched off. Figure 9.29 shows how fast the water in tank A cools down after the heater is switched off.

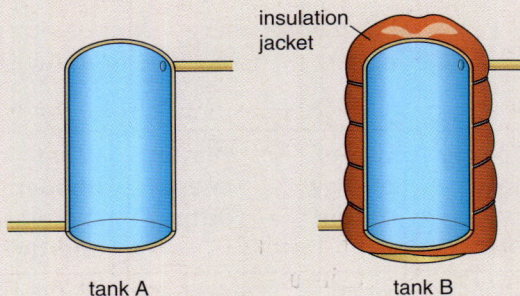

Figure 9.28

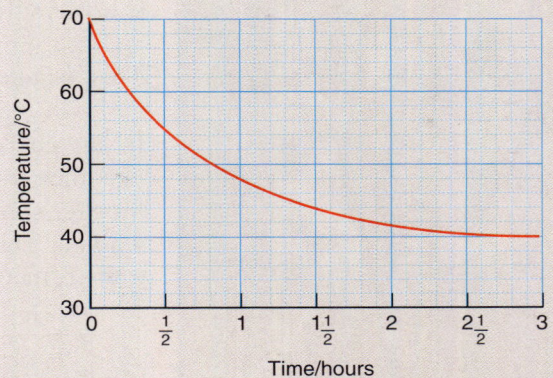

Figure 9.29

a) In which tank will the water cool down faster?
b) Copy and complete the following sentence by choosing the correct phrase.
  After three hours, the water in tank A will be *colder than / the same as / hotter than* the water in tank B.
c) Carefully copy Figure 9.29. Draw a second line on your graph to show how the water in tank B is likely to cool down.

❶❻ Write down six ways to reduce the amount of energy used in a home.

9.5

Energy efficiency

Energy from nothing!

We often hear that the Earth's energy resources are running out. This is because most of them are non-renewable resources such as fossil fuels. At the same time, we are encouraged to use more renewable energy resources like wind power and water power. It would be wonderful if we could make energy from nothing. But this is impossible!

Energy can be **transferred** (moved) from one place to another. Energy can be **transformed** (changed) from one form another. But energy cannot be created or destroyed. In other words, we can't make energy and it can't simply disappear. The total amount of energy you start with is always the same as the total amount of energy you end up with.

This is called the **law of conservation of energy**.

In any energy transfer or transformation, the energy spreads out and becomes less useful to us. All devices waste some energy. Devices transfer some energy to where it is wanted and in the form that is wanted. This is useful energy transfer. The rest of the energy is transformed into forms that are not wanted. This energy is wasted.

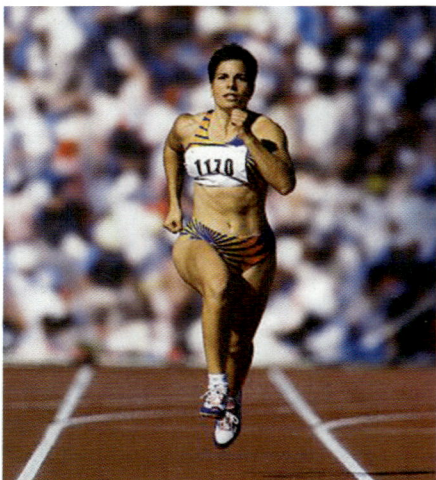

Figure 9.30 Chemical energy in the food of the sprinter changes to kinetic (movement) energy, heat in her muscles and a small amount of sound energy.

Figure 9.31 A firework transforms (changes) chemical energy into light, heat, sound and gravitational potential energy.

17 Copy and complete the following sentence using the words in this box.

| conservation | wasted | total |

The law of _____ of energy says that the useful energy output from a machine plus the _____ energy must equal the _____ energy supplied to the machine.

18 The devices listed in Table 9.4 transform energy from one form into other forms. Copy and complete the table. The first row has been done for you.

Device	Energy input	Useful energy output	Wasted energy output	What happens to the wasted energy?
Motorbike engine	Chemical	Kinetic (movement)	Heat and sound	Warms the air
Firework				
Electric motor				
Coal-burning power station				

Table 9.4

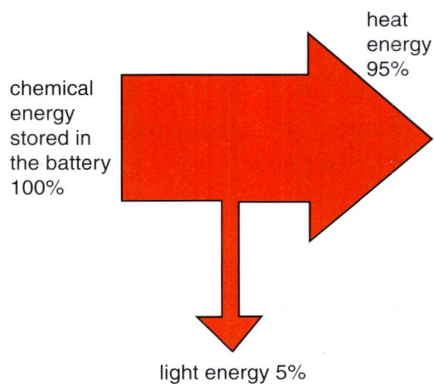

Figure 9.32 A Sankey diagram for a torch

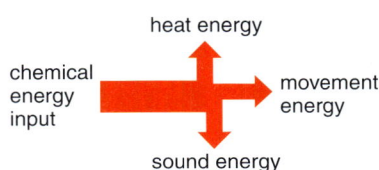

Figure 9.33 A Sankey diagram for a motorbike engine

The **efficiency** of a machine is the proportion of the energy supplied that is transferred in a useful form.

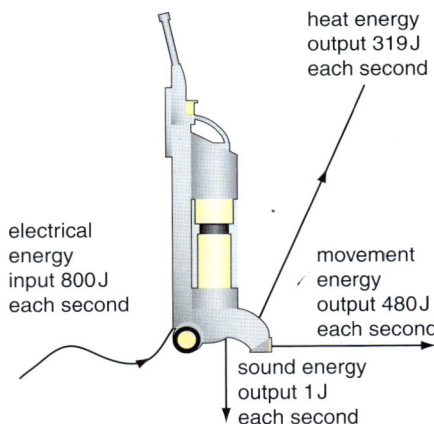

Figure 9.34 The energy transferred by a vacuum cleaner

Figure 9.32 is a **Sankey diagram**. It can be used to show how a device transforms energy from one form into another. The wider the arrow the greater the energy transformed. Figure 9.32 shows that the chemical energy stored in the torch battery is transformed into light and heat. It also shows that the total amount of energy stays the same. The total energy has been **conserved**. But we only want the light energy. The heat energy is wasted.

A motorbike engine transforms chemical energy from the petrol into kinetic (movement) energy (Figure 9.33). But it also transfers a lot of energy into heat and sound. This energy is wasted.

In most energy transfers, energy is transferred to the air. This causes the air to become warmer. Because the energy spreads out, any increase in temperature is usually too small to notice. So the energy that started in a useful form ends up wasted but it has not disappeared.

In any energy transfer or transformation, the energy spreads out and becomes less useful to us.

How good is a device or machine at transferring energy in a useful way?

The **efficiency** of a device tells us how good it is at transferring energy in useful forms. A machine is 100% efficient if the total energy going in is the same as the useful energy going out. No machine can be more than 100% efficient.

The efficiency of a device can be worked out using this equation:

$$\text{efficiency} = \frac{\text{useful energy transferred by the device}}{\text{total energy supplied to the device}}$$

Example
A vacuum cleaner (Figure 9.34) transforms electrical energy into kinetic energy. This is useful energy. But it also transfers energy as heat and sound. This energy is not useful; it is wasted.

What is the efficiency of the vacuum cleaner?

$$\text{efficiency} = \frac{\text{useful energy transferred by the device}}{\text{total energy supplied to the device}}$$

The device in this case is the vacuum cleaner.

$$\text{efficiency} = \frac{480}{800} = 0.6$$

Sometimes, efficiency is written as a percentage. So, in this case, it would be:

$$\frac{480}{800} \times 100 = 60\%$$

The bigger the percentage of energy usefully transferred by a device, the more efficient the device is.

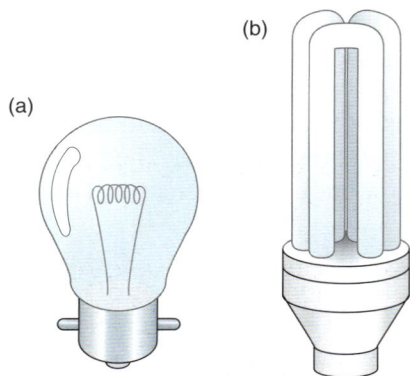

Figure 9.35 a) A filament light bulb
b) A low-energy light bulb

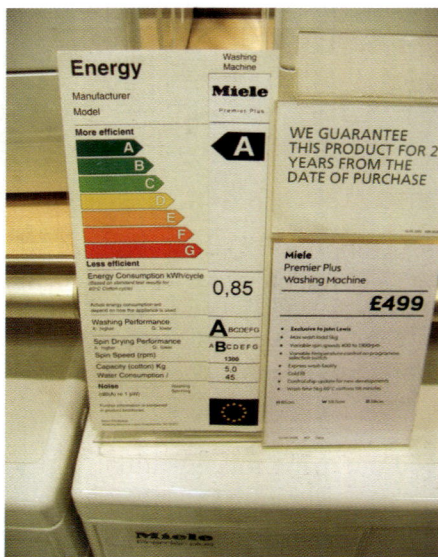

Figure 9.36 All new fridge freezers (and other electrical devices) must have a European Energy Label. Information on the label compares the efficiency and running costs of different models.

Low-energy light bulbs

An ordinary filament lamp is about 5% efficient. It transforms most of the electrical energy into heat. This energy is wasted.

Low-energy light bulbs are much more efficient. They transform a larger percentage of the electrical energy into the light energy that we want. This means that a 20 watt low-energy bulb can give out as much light as a 100 watt filament lamp. (See Section 11.3.)

19 A diesel engine transforms 40% of the input energy to kinetic energy, 15% to sound and 45% to heat.
   a) Draw a Sankey diagram for the diesel engine.
   b) What is the efficiency of the engine?

20 A car engine wastes a lot of energy as heat. How could some of this wasted heat be usefully used?

21 When Steve and Sue go out for the evening, Sue always switches the lights and TV off. She says that it saves energy. But, Steve is puzzled. He knows that energy cannot be destroyed. Therefore it must always be there and cannot be saved. Explain carefully, using the idea of energy conservation, what Sue means by 'saving energy'.

22 An electric oven transforms electrical energy into heat. The electric oven is 70% efficient.
   a) What percentage of the electrical energy is usefully transferred as heat?
   b) What percentage of the electrical energy is wasted?
   c) Draw a Sankey diagram for the electric oven.

23 The Sankey diagram in Figure 9.37 shows that a street lamp transforms electrical energy to light and heat. What is the efficiency of the street lamp?

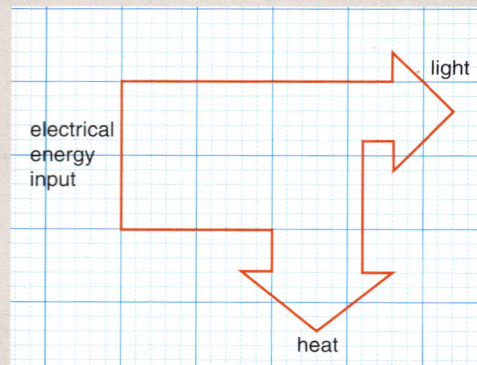

Figure 9.37

Activity – Investigating the efficiency of an electric motor

When an electric motor is used to lift a weight it transforms electrical energy into useful gravitational potential energy. Susan investigated how the efficiency of this energy transfer changes as heavier weights are lifted. She used the apparatus shown in Figure 9.38.

Susan used a joulemeter to measure the electrical energy supplied to the motor each time it lifted a weight 0.82 m. She repeated the experiment using different weights. For each weight she took three measurements of the electrical energy supplied and recorded the average. Her results are shown in Table 9.5.

Electrical energy input in joules	Weight lifted in newtons	Calculated efficiency
15	2	10.7%
17	3	14.2%
21	4	15.3%
26	5	15.5%
33	6	14.6%
46	7	12.2%

Table 9.5

❶ Copy and complete the following sentences using words from this box.

> controlled   dependent   independent
> reliable   seven   newtons

Susan used a range of weights from two _____ to _____ newtons. Changing the weight caused the energy input to the motor to change. This means that weight is the _____ variable and energy input is the _____ variable. Each weight was lifted exactly 0.82 m. This means that the distance the weight was lifted is a _____ variable. Taking three measurements of the electrical energy for each weight made the average value recorded in the table more _____.

Remember, the **independent variable** is what is changed. The **dependent variable** is what changes because of this.

❷ Draw a graph of the weight lifted (horizontal axis) against efficiency (vertical axis).
❸ How does the efficiency of this motor change as heavier weights are lifted?

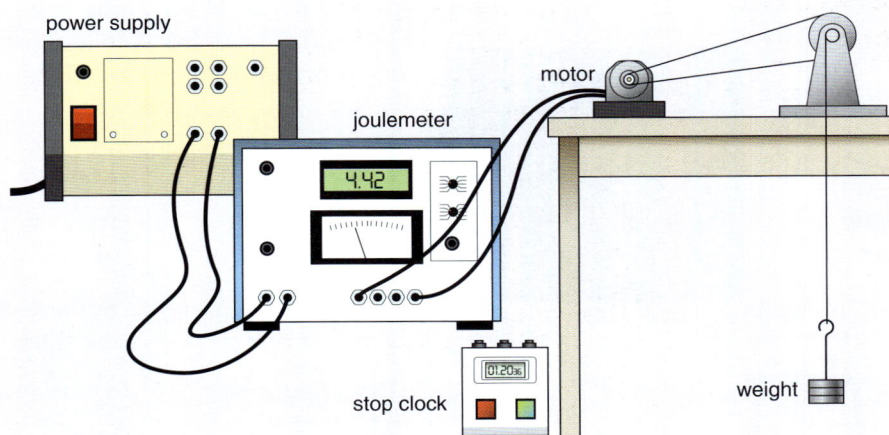

Figure 9.38 Apparatus used to investigate the efficiency of an electric motor

Summary

✓ **Infra-red waves** are also called **thermal radiation**. They can travel through a vacuum.

✓ All objects **emit** (give out) and **absorb** (take in) thermal radiation.

✓ Dark-coloured matt surfaces are good emitters and good absorbers of thermal radiation.

✓ Light-coloured surfaces are poor emitters and poor absorbers of thermal radiation.

✓ Shiny surfaces reflect thermal radiation better than dull matt surfaces.

✓ **Conduction** and **convection** involve the movement of particles.

✓ A material that contains mobile electrons is a good conductor of heat.

✓ The rate at which an object loses heat depends on:
  ● the material the object is made from;
  ● the shape of the object;
  ● the size of the object;
  ● the difference in temperature between the object and its surroundings.

✓ Energy can be **transferred** (moved) from one place to another. Energy can be **transformed** (changed) from one form to another.

✓ **The law of conservation of energy** says: 'Energy cannot be created or destroyed. It can only be transformed from one form to another form.'

✓ When energy is transferred or transformed some energy is always wasted.

✓ Wasted energy spreads out and becomes less useful.

✓ The **efficiency** of a device is the proportion of the energy input that is transferred in a useful form.

✓ $$\text{efficiency} = \frac{\text{useful energy transferred by the device}}{\text{total energy supplied to the device}}$$

EXAMQUESTIONS

❶ Figure 9.39 shows three examples of heat transfer. Match each picture to the correct heat transfer process: conduction, convection or radiation. *(3 marks)*

a)

b)

c)

Figure 9.39 a) A spoon gets warm in a bowl of hot soup
b) Heat reaches the Earth from distant stars
c) Water rises when it is heated in a flask

❷ List A below gives different methods of insulating a home. List B gives information about the methods of insulation. Copy list A and list B. Draw a line from each method of insulation in list A to a correct piece of information in list B. All four boxes in both lists must be used.

List A	List B
Carpet	Air is trapped between two pieces of glass
Loft insulation	Reduces heat loss through the floor
Double glazing	Traps small pockets of air
Cavity insulation	Is often used to fill the gap between the inside and outside walls of a house

(3 marks)

❸ Marion and Jim have been asked to find out which of three materials, K, L or M, would be the best for making a winter coat. They each produce a plan.

Marion's plan
1 From each material, cut a rectangle 8 cm by 20 cm.
2 Wrap material K round an empty metal can, hold it in place with an elastic band.
3 Pour 200 cm^3 of boiling water into the can; place a thermometer in the water.
4 Wait until the temperature reaches 85 °C then start a stop watch.
5 Take the temperature every minute for 10 minutes.
6 At the end of 10 minutes throw the water away.
7 Repeat the experiment with material L, then material M.
8 Each time use 200 cm^3 of water and wait until the temperature reaches 85 °C before starting the stop watch.

Jim's plan
1 Cut a piece of material from each of K, L and M.
2 Wrap K around an empty metal can.
3 Pour in some hot water.
4 Start a stop watch and take the temperature every few minutes.
5 When the water has cooled down throw it away.
6 Repeat the experiment with the other two materials.
   a) Give four reasons why Marion's results will be more useful than Jim's results.
                                              (4 marks)
   b) Marion's results are presented in the graph in Figure 9.40.

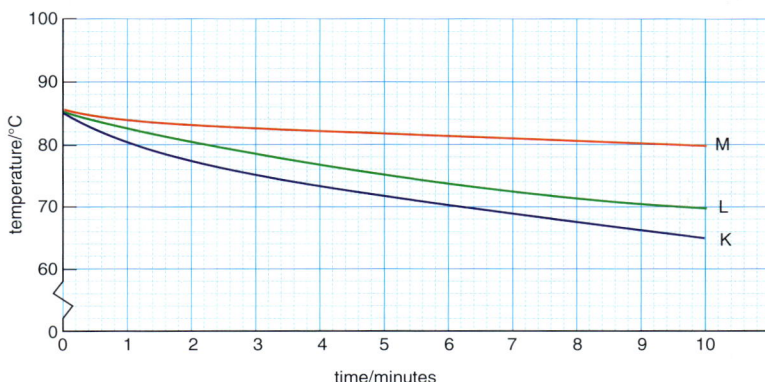

Figure 9.40 A graph of Marion's results

Which of the materials K, L or M should the winter coat be made from? Give a reason for your answer.   *(2 marks)*

❹ The Sankey diagram in Figure 9.41 shows what happens to the input energy for a television.

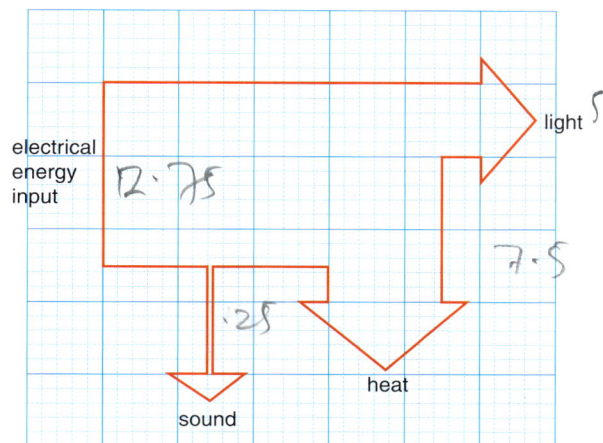

Figure 9.41

a) Use the following equation to calculate the efficiency of the TV. Show clearly how you work out your answer.

efficiency = $\dfrac{\text{useful energy transferred by a device}}{\text{total energy supplied to device}}$

                                              (2 marks)

b) Joe was arguing with a friend about conservation of energy. Joe said 'As energy is conserved, it must always be there. So there is no point in switching things like a TV or computer off.' Explain why Joe is wrong.   *(2 marks)*

EXAMQUESTIONS

Chapter 10
How do we generate and use electricity?

At the end of this chapter you should:

✓ know that electrical appliances transform (change) electrical energy to other forms of energy at the flick of a switch;

✓ know that the faster an appliance transforms energy, the higher the power of the appliance;

✓ understand how the National Grid transfers electricity from power stations to consumers;

✓ know how electricity is generated in power stations;

✓ know that electricity is generated using a variety of energy sources;

✓ know the advantages and disadvantages of different energy sources for generating electricity;

✓ be able to calculate the amount of electrical energy transferred from the mains supply, and the cost of this energy.

Figure 10.1 Different energy resources are used to generate electricity. But in the UK, most electricity is still generated from fossil fuels in large power stations.

10.1 # Why is electrical energy so useful?

In most homes there are different devices, such as hairdryers, washing machines, TVs and lamps. Although they are all different, they all work by transforming (changing) electrical energy into other forms of energy. Devices that change electrical energy into other forms of energy are also called electrical **appliances**.

Electrical energy is useful because it is easily transformed (changed) into other forms of energy, such as:
- heat (thermal energy);
- light;
- sound;
- kinetic (movement) energy;
- gravitational potential energy.

This is why we have so many appliances that work from the electricity mains supply.

Figure 10.2 shows five electrical appliances and the energy transformations (changes) they make.

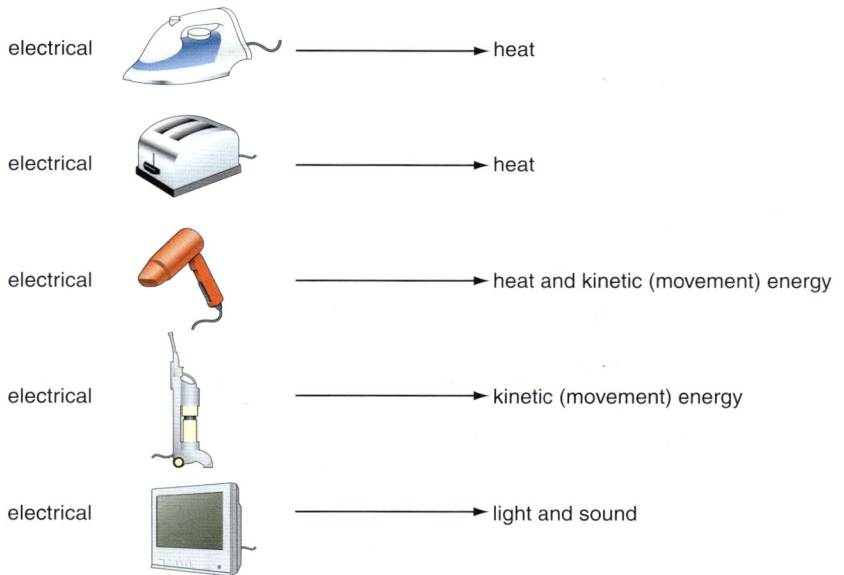

❶ Copy and complete Table 10.1. The first line has been done for you.

Electrical device	Transforms electrical energy to:
Drill	Kinetic (movement) energy
MP3 player	
Hotplate	
Toothbrush	

Table 10.1

Figure 10.3 Transforming electrical energy into useful forms of energy

electrical ⟶ heat

electrical ⟶ heat

electrical ⟶ heat and kinetic (movement) energy

electrical ⟶ kinetic (movement) energy

electrical ⟶ light and sound

Figure 10.2 Some everyday electrical appliances and the energy changes they make

Figure 10.4 There are lots of electrical appliances in a kitchen.

❷ Look at Figure 10.4. Write down a list of the appliances in Figure 10.4 that are designed to transform electrical energy into heat.

❸ The builders in Figure 10.5 are using electricity to make their job easier.

Copy and complete the sentences below using the words in this box.

| electrical | energy | kinetic | move |
| | potential | | |

The petrol generator transfers chemical _____ from the petrol into _____ energy. The electric motor makes the conveyor belt _____. When it is working the conveyor belt has _____ energy.

The bags of cement moving up the conveyor belt gain gravitational _____ energy.

Figure 10.5

Electrical energy and power

The amount of electrical energy that a device transforms depends on two things:
- how long the device is used;
- the rate at which the device transforms energy (uses electricity).

The rate at which a device transforms energy is called its **power**.

Power is measured in watts (W) or joules per second.

1 watt = 1 joule per second (1W = 1 J/s)

Appliance	Power rating
Lamp	60 W
Television	150 W
Toaster	250 W
Vacuum cleaner	770 W
Hairdryer	1200 W
Iron	1800 W

Table 10.2 The power rating of some everyday electrical devices

4 Each of the following appliances is used for 15 minutes.

1200 W	hairdryer
100 W	light bulb
20 W	radio
900 W	vacuum cleaner

a) Which appliance transfers the most energy?

b) Which appliance transfers the least energy?

Give a reason for each of your choices.

A device that transforms 1 joule of energy every second from one form to another has a power of 1 watt. So a 350 watt electric saw transforms 350 joules of electrical energy to other forms of energy every second it is switched on.

Figure 10.6 An electric saw

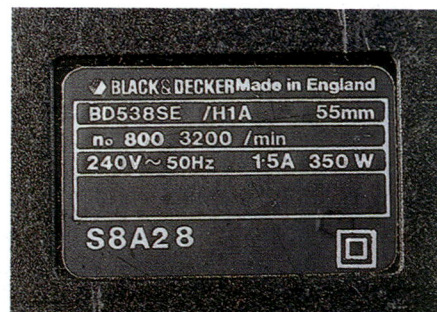

Figure 10.7 The information plate shows the power of the electric saw.

Power is also measured in kilowatts (kW).

1 kilowatt (kW) = 1000 watts (W)

Table 10.2 shows the power of some of the devices mentioned earlier in this section.

Activity – Using electrical devices at home

Think about your own home and all the electrical appliances that you use regularly.

1 Make a list of all the appliances in your home that work from the mains electricity supply.

2 Figure 10.8 shows two different electric fans. Fan A works from the 240 V mains electricity supply. Fan B is a 3V battery-operated fan.

Figure 10.8 a) Fan A – mains operated fan b) Fan B – battery operated fan

Copy and complete Table 10.3. You can put more than one advantage or disadvantage in each space if you wish. One advantage has been written in for you.

	Advantage	Disadvantage
Fan A (mains operated)	It moves a lot of air every second	
Fan B (battery operated)		

Table 10.3

10.2

Paying for electricity

Every time an appliance is used, energy is taken from the mains supply. The person paying the electricity bill pays for this energy.

⑤ Table 10.4 shows the powers of four different appliances.

Appliance	Power
Toaster	400 W
Microwave	600 W
Hairdryer	1.2 kW
Fan	60 W

Table 10.4

Which appliance costs the most to use for one hour? Give a reason for your answer.

⑥ Calculate the cost of using the following appliances. Each kilowatt-hour of energy costs 12p.
 a) A 2 kW heater used for 4 hours.
 b) A 4 kW cooker used for 2 hours.
 c) A 1500 W hairdryer used for 15 minutes (¼ hour).
 d) A 100 W light bulb used for 5 hours.

Appliances with a high power rating (see Section 10.1) take a lot of energy from the mains supply. For example, an electric shower with a power rating of 9000 watts takes 9000 joules of electrical energy every second when it is switched on. If your shower takes 5 minutes (300 seconds) that's a lot of energy! So any high-power appliance costs a lot to use.

The cost also depends on how long the appliance is used. The longer the appliance is on, the more it costs. Spend twice as long in the shower and it costs twice as much.

Energy is usually measured in joules. But one joule of energy is not much. Because of this, the electricity companies measure energy in a much larger unit. The unit they use is the **kilowatt-hour (kWh)**.

Working out the energy in kilowatt-hours

To work out the energy transferred by an appliance you use the equation:

$$\text{energy transferred} = \text{power} \times \text{time}$$
$$\text{(kilowatt-hours)} \quad \text{(kilowatts)} \quad \text{(hours)}$$

Working out the bill

The cost of using mains electricity is worked out by multiplying the number of kilowatt-hours of energy used by the cost of one kilowatt-hour.

$$\text{total cost} = \text{number of kilowatt-hours} \times \text{cost per kilowatt-hour}$$

Reading the electricity meter

Somewhere in your home there will be an electricity meter. This measures all the electrical energy taken from the mains supply and used in your home. The more energy you take, the higher the bill.

Figure 10.9 shows the electricity meter in Dan's house. The top picture shows the meter in February. The bottom picture shows the meter three months later in May. The readings on the meter show the number of kilowatt-hours of energy used.

The electricity company uses the readings to work out how much money Dan has to pay.

7 0 2 2 2
kW h

11 February

7 0 4 6 9
kW h

13 May

Figure 10.9 Dan's electricity meters, showing readings three months apart

7 Figure 10.11 shows a graph of the readings taken from Ashfaq's electricity meter over one year.

a) Between which two months does Ashfaq use the smallest amount of electrical energy?

b) Why is the graph line steepest between November and February?

c) How can long-term weather forecasts help power stations to predict the amount of electricity needed?

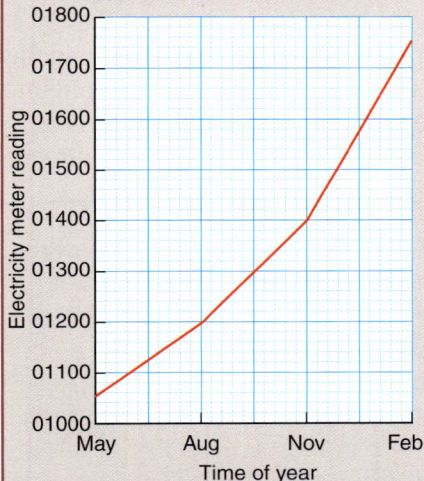

Figure 10.11 A graph of the readings taken from Ashfaq's house

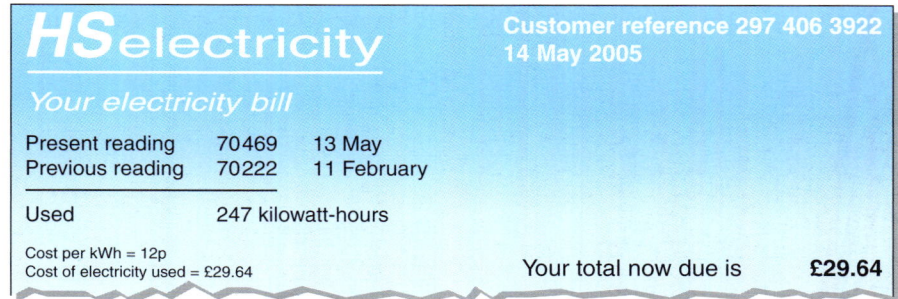

HSelectricity

Customer reference 297 406 3922
14 May 2005

Your electricity bill

| Present reading | 70469 | 13 May |
| Previous reading | 70222 | 11 February |

| Used | 247 kilowatt-hours |

Cost per kWh = 12p
Cost of electricity used = £29.64

Your total now due is **£29.64**

Figure 10.10 Dan's electricity bill

Between February and May, Dan used

70 469 − 70 222 kilowatt-hours of electrical energy

This means he used 247 kilowatt-hours of electrical energy. Each kilowatt-hour costs 12p.

Total cost = number of kilowatt-hours × cost per kilowatt-hour
= 247 × 12p = 2964p = £29.64

8 Sue gets home to find that her dog Zach has eaten part of the electricity bill.
How much does Sue need to pay the electricity company?

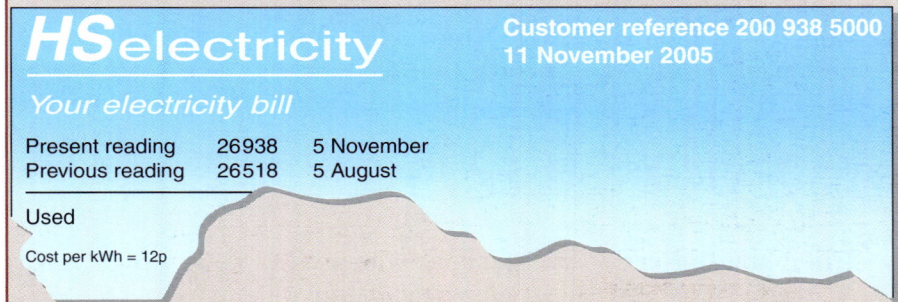

HSelectricity

Customer reference 200 938 5000
11 November 2005

Your electricity bill

| Present reading | 26938 | 5 November |
| Previous reading | 26518 | 5 August |

| Used | |

Cost per kWh = 12p

Figure 10.12 All that is left of Sue's electricity bill

10.3 # The National Grid

Electricity is generated in power stations. It is then sent through a network of cables and transformers to our homes, offices, schools and factories. This network is called the **National Grid**. The people who use the electrical energy are called consumers.

The National Grid links all the main power stations and smaller electricity generators together. This means that people still get the electricity they want even when some power stations are shut down. Some power stations can supply extra power to the grid when the engineers know there will be a high demand. (Imagine how much extra power is needed at half-time in a televised world cup football match. Everyone is switching on electric kettles for cups of tea or coffee at the same time!)

Figure 10.13 A simplified diagram of the National Grid network

Every second, huge amounts of electrical energy are sent through the National Grid from power stations to consumers. This energy is sent at high voltage. To do this, a step-up **transformer** increases the voltage in the electricity from the power station. When the voltage goes up, the current in the cables goes down. If the current is smaller, the heating in the wires is also smaller. This means that less electrical energy is changed into heat and wasted. So using a high voltage makes the National Grid more efficient. More of the energy from the power station gets to the consumer.

Figure 10.14 Pylons carry the main cables of the National Grid to all parts of the country.

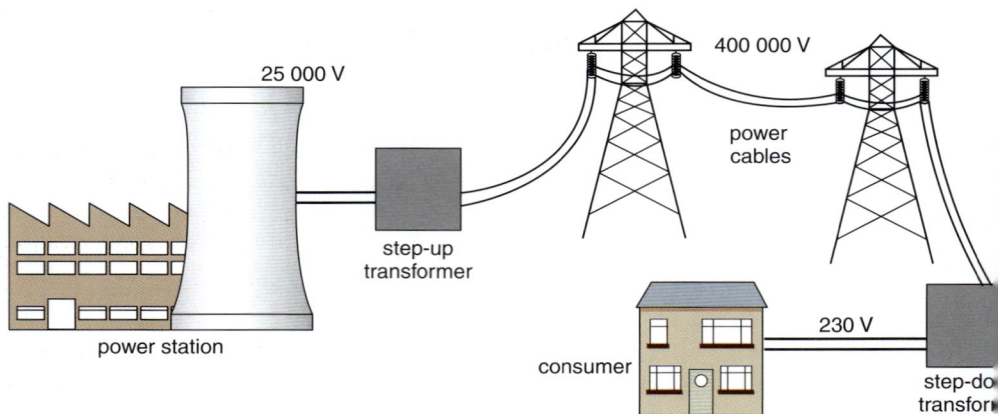

9 Copy list A and list B below. Draw a line from each name in list A to its correct meaning or use in list B.

List A	List B
step-up transformer	someone who uses electrical energy
National Grid	used to increase the voltage
consumer	where electricity is generated
power station	system of cables and transformers

10 Transformers are not 100% efficient. They waste some energy as heat and sound.
   a) What do we mean when we say something is 100% efficient?
   b) What does the waste energy from a transformer do to the air around it?

11 Look at Figure 10.13.
   a) What is the voltage used in the National Grid power cables?
   b) Why are the power cables held high above the ground?
   c) What is the voltage used by consumers in their homes?
   d) Why must the voltage used in homes be much lower than the voltage in the National Grid cables?
   e) Copy and complete the following sentence by choosing the correct line from the box.

Using a very high voltage in the power cables

increases
makes no difference to
decreases

the efficiency of the National Grid.

10.4 How should we generate the electricity we need?

Electricity has to be generated using another source of energy. In Britain most electricity is generated using the energy from coal, oil and natural gas. These are fossil fuels. When they burn, fossil fuels transfer energy as heat. Fossil fuels are **non-renewable energy sources**. Once they are gone, they are gone for ever. They can't be renewed.

Fossil fuels will eventually run out. Figure 10.15 shows how long we can expect them to last if we continue to use their reserves at the present rate.

If we want fossil fuels to last longer than this, we must learn to use them more efficiently. This is very important for oil and natural gas.

Energy sources like the wind and the tides will never run out. They are replaced faster than we can use them. We call them **renewable energy sources**.

Thermal power stations

A thermal power station uses the heat from a fuel to generate electricity.

Coal or oil is burned to heat water and produce steam. The steam is made to drive turbines. The turbines turn generators. The generators produce electricity (Figure 10.16).

In a gas-burning power station there is no need to produce steam. The burning gas produces fast-moving hot air, which drives the turbine directly.

Nuclear power stations generate electricity in a similar way to power stations that burn coal. But the fuel, which is mainly uranium or plutonium, is not burned. The heat needed to make the steam is given out when the uranium or plutonium atoms break up inside the nuclear reactor (Figure 10.17).

The break-up (splitting) of large atoms in elements like uranium and plutonium is called **nuclear fission**. When nuclear fission happens, energy is released.

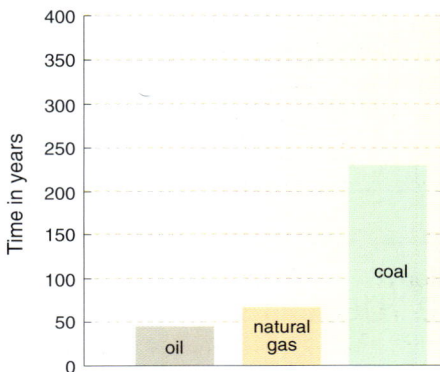

Figure 10.15 How long will fossil fuels last?

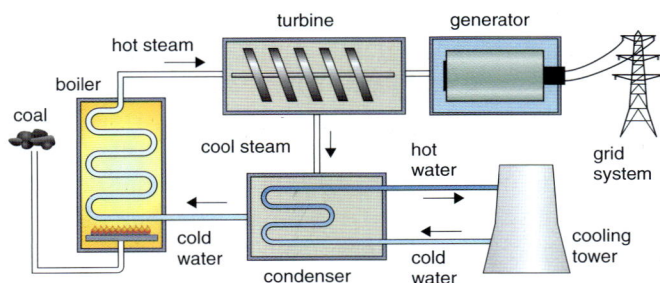

Figure 10.16 Using coal or oil to generate electricity

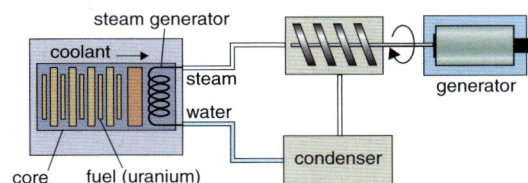

Figure 10.17 Using nuclear fuel to generate electricity

12 Figure 10.18 shows how the energy from coal can be transformed into electrical energy. Copy and complete Figure 10.19 by filling in the missing words.

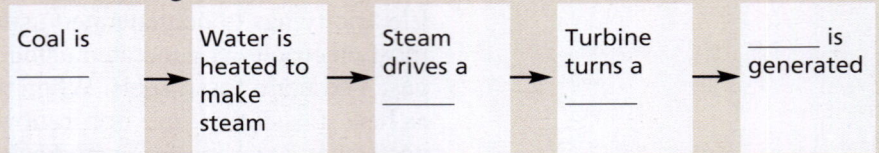

| Coal is _____ | → | Water is heated to make steam | → | Steam drives a _____ | → | Turbine turns a _____ | → | _____ is generated |

Figure 10.18 How energy from coal is transformed into electrical energy

13 Each of the statements A, B, C and D is about an energy source. Copy and complete Table 10.5 by putting each statement in one of the columns.
A These energy sources will run out.
B This can come from the Sun.
C These energy sources will never run out.
D Oil was made millions of years ago.

Non-renewable energy source	Renewable energy source

Table 10.5

10.5 Why are non-renewable fuels used to generate electricity?

Thermal power stations can generate electricity at any time. It doesn't matter if it's night or day, summer or winter. As long as the fuel keeps coming, the power station keeps generating. This makes non-renewable fuels reliable energy sources.

Coal-burning power stations have power outputs of about 2000 megawatts (that's 2000 million watts). So a few power stations can provide the power needed by millions of consumers.

Sometimes power stations are closed for maintenance. Later the power station needs to be started up again. The time it takes to get started again is different for different types of power station. If possible, coal-burning and oil-burning power stations are kept running all the time. Allowing the furnaces to cool down is likely to damage them.

Nuclear fuel is a concentrated energy source. One kilogram of nuclear fuel gives out the same amount of energy as 20 000 kilograms of coal. But nuclear power stations are very expensive to build. They are also very expensive to decommission (take to pieces) safely when they aren't needed any more.

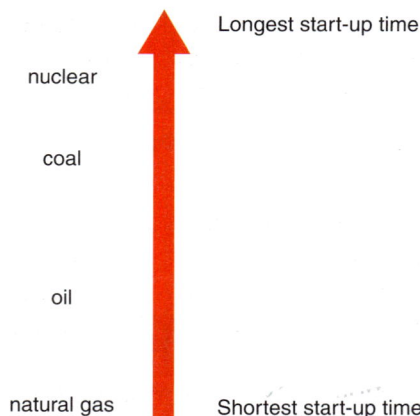

Longest start-up time
nuclear
coal
oil
natural gas   Shortest start-up time

Figure 10.19 Some types of power station can start generating faster than others.

How does the use of non-renewable fuels affect the environment?

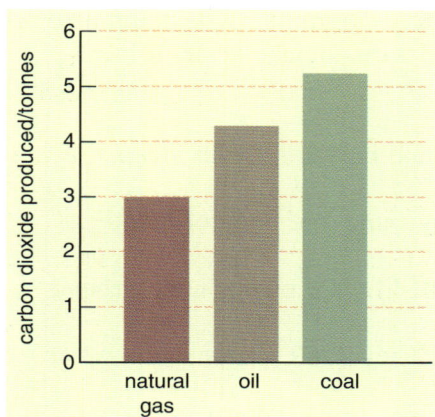

Fossil fuels and global warming

Power stations that burn fuels produce vast quantities of carbon dioxide (Figure 10.20). So, as more fossil fuels are burned, more carbon dioxide is put into the atmosphere. Many people think that this extra carbon dioxide has led to more heat being trapped in the atmosphere. And this has led to a worldwide increase in temperature. This increase in worldwide temperature is called **global warming** (Section 6.4).

Fossil fuels and acid rain

When coal and oil burn in power stations, a lot of sulfur dioxide is produced as well as carbon dioxide. If the sulfur dioxide gets into the atmosphere and dissolves in rain, it forms an acidic solution. This is called **acid rain** (Section 6.4). Natural gas does not produce sulfur dioxide when it burns, so gas-fired power stations don't cause this problem.

Acid rain from coal-burning and oil-burning power stations can be reduced. But this is expensive.
- Either the sulfur can be removed from the fuels before they are burned.
- Or the sulfur dioxide can be removed from the waste gases.

Both these methods of reducing acid rain are used in power stations, but they increase the cost of electricity.

Into the future

Materials such as palm nuts and straw can be used as **bio-fuels**. They emit carbon dioxide when they burn, but they also use up carbon dioxide as they grow. So, overall, they put no extra carbon dioxide into the atmosphere. Some power stations are now trying blends of bio-fuel and coal. Elean Power Station in Cambridgeshire generates 36 MW of power by burning straw. One electricity company, in Queensland, Australia, generates enough electricity for 1200 homes by burning nut shells.

Nuclear fuels and the environment

Nuclear fuels do not produce carbon dioxide or sulfur dioxide. So nuclear power stations do not add to the problems of global warming or acid rain. But nuclear power stations do produce radioactive waste. This must be stored safely for a long time, sometimes for thousands of years.

Some people worry that nuclear power stations may leak radiation. But when nuclear power stations are working properly, no radiation should enter the environment.

Figure 10.20 The amount of carbon dioxide produced for each kilowatt of power generated by burning a fossil fuel

⓮ Read the following statements. Then choose the three statements that give a reason for using coal to generate electricity.
  A Burning coal produces carbon dioxide and sulfur dioxide.
  B Coal-burning power stations can generate electricity all year round.
  C One coal-burning power station can generate electricity for a million people.
  D Coal is a non-renewable fuel.
  E There is enough coal to last hundreds of years.
  F Coal-burning power stations are about 30% efficient.

⓯ What do most people think is the cause of global warming?

⓰ We cannot generate our electricity from fossil fuels for ever. Why not?

17 A nuclear power station produces 2 400 000 kW h of electrical energy every hour. 1 kg of nuclear fuel produces 1 600 000 kW h of electrical energy.
How much nuclear fuel does the power station use in one hour?

18 Richard may buy some land to build a house on. The land is close to a nuclear power station. What evidence could a health physicist working at the power station give Richard to persuade him that it is safe to live in a house built on the land?

Activity – Nuclear waste? Not in our backyard!

Suppose you are the scientist on a safety committee. The committee must recommend the best way to deal with nuclear waste. It has come up with three options.

Option 1: Seal the waste in concrete and bury it for ever, deep underground.
Option 2: Bury the waste deep underground, but monitor it and recover it if necessary.
Option 3: Bury waste with a short half-life 100 metres underground.

Some people have started protesting. They don't want nuclear waste buried near their homes.

1 Why don't the protesters want nuclear waste buried near their homes?
2 You are the scientist on the committee. What could you say to persuade the protesters that burying nuclear waste is safe?
3 Explain why the committee has said that waste with a short half-life can be buried only 100 metres underground.

Section 11.11 will help you with this activity.

19 How many 4 MW (megawatt) wind turbines are needed to replace a 2000 MW coal-burning power station?

20 Copy and complete the following sentence.

Gas power stations start generating electricity _____ than coal power stations because they do not heat _____ and change it into steam.

10.6 Why use renewable energy sources to generate electricity?

Non-renewable fuels like oil and gas will not last for ever. They are running out and more and more electricity is needed every year.

At the moment Britain produces only 3% of its electricity from renewable energy sources. The aim is to increase this to 25% by the year 2025. So how will this be done?

Using the tides

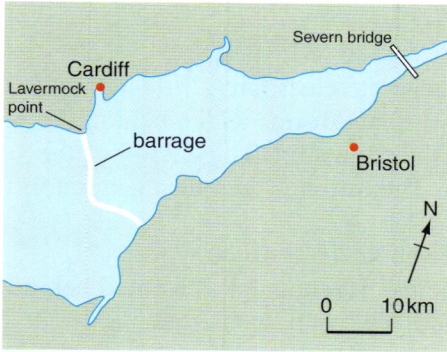

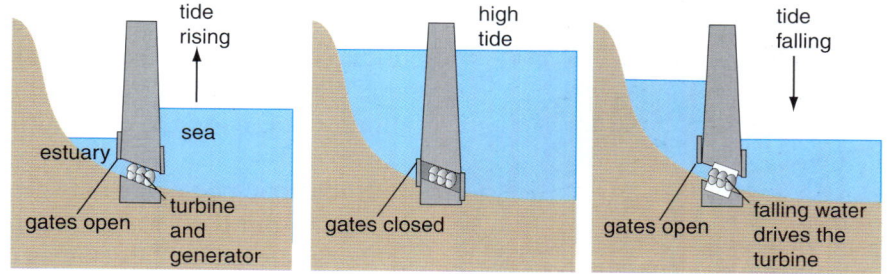

Figure 10.22 A tidal barrage across the River Severn could supply as much as 6% of Britain's electricity.

Figure 10.21 A tidal barrage

Every day the tides rise and fall. Massive amounts of water move in and out of river estuaries. Using energy from the tides it might be possible to generate up to 20% of the electricity we need in Britain.

A tidal barrage, which is like a dam, is built across a river estuary (Figure 10.21). When the tide comes in, water is trapped behind the barrage. When the tide goes out, underwater gates are opened. Water rushes through these gates, driving turbines. The turbines turn generators that produce electricity.

A tidal generator cannot produce electricity all of the time. But we know when the tides will rise and fall, so the output from a tidal generator is reliable and there is no pollution.

Tidal barrages must be built across river estuaries (Figure 10.22). Because of this, they can disturb the flow of the river. This may destroy the habitats of wading birds and the mud-living organisms on which they feed.

Using waves

Britain is surrounded by the sea and its waves. So why are there only a few small wave generators? The answer is simple. It is difficult to capture the energy from waves and transform it into large amounts of electricity. As well as this, a wave generator must be extremely tough to survive the pounding from rough seas during storms.

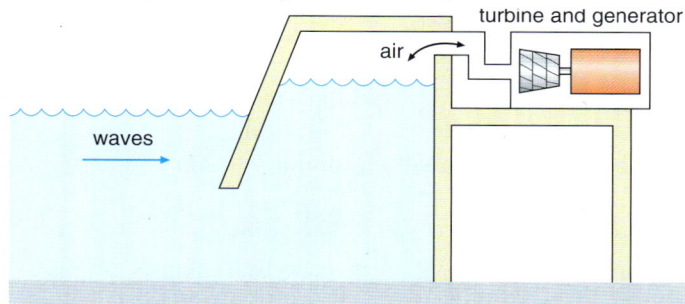

Figure 10.23 LIMPET is the world's first commercial wave generator. It has been generating electricity since November 2000.

Figure 10.24 A wave generator

LIMPET is a wave generator (Figure 10.23), built on the Scottish island of Islay. The strong movement of the Atlantic waves forces air to drive a turbine. The turbine then turns a generator (Figure 10.24).

Using hydroelectric power

The energy of a river can be used to generate electricity. First a dam is built across the river. This traps the water and forms a lake behind the dam. When the water is released it rushes downhill. The energy of the falling water is used to drive turbines. The turbines then turn generators.

Hydroelectric power stations generate about 10% of the world's electricity. But they bring some problems. The new lakes formed behind the dams often flood large areas of land. This may mean that:
- forests are cut down;
- farmland is lost;
- wildlife habitats are destroyed.

In China, the world's largest power project involved building a dam across the Yangtze River (Chang Jiang in Chinese). The lake behind the dam flooded so much land that about 1.5 million people had to be moved to new homes.

Supply and demand

The demand for electricity changes during the day. When demand is high, a hydroelectric pumped storage system can provide the extra electricity needed.

In only a few seconds, water in the top lake can be released. As the water falls to the bottom lake it drives a turbine. The turbine then turns a generator.

At night, surplus electricity is used to pump water back into the top lake. So the energy is stored in the water rather than being wasted. The power station is now ready to satisfy higher demands again.

Figure 10.25 Hydroelectric power stations can be very large in size and generate vast amounts of power.

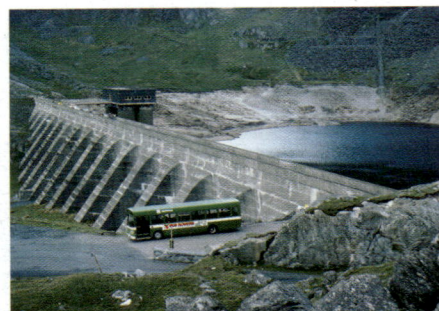

Figure 10.26 Ffestiniog Power Station was the first major pumped storage power station in the UK.

㉑ Copy and complete the sentences below using the words in this box.

barrage	estuary	falling	flood
fuel	generator	pollution	
predictable	renewable	under	very

a) To generate electricity using energy in the tides, a large tidal _____ must be built across a river _____.Tides are _____ and produce no _____.

b) A wave generator does not use a _____.

Waves are a _____ energy resource. Wave power stations are built where there are _____ strong waves. In the future most wave generators may be _____ the seas.

c) Hydroelectric power stations use _____ water to drive a turbine. The turbine turns a _____. The lake formed behind the dam may _____ a large area of land.

Activity – Investigating a water-powered electricity generator

Anya investigated the model generator shown in Figure 10.27. Water from a tap drives the turbine. The turbine turns the generator. The electricity generated lights a bulb.

Anya wanted to find out how the volume of water hitting the turbine each second changed the output voltage of the generator.

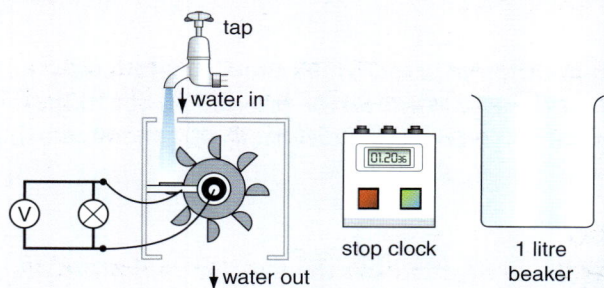

Figure 10.27 The apparatus used by Anya to investigate the generator

	Volume of water collected
1st try	42 ml
2nd try	52 ml
3rd try	68 ml

Table 10.6 Anya's first set of results

Anya put the turbine under the tap. She slowly increased the flow of water until the turbine just started to turn. Anya then placed a beaker into the water flow, timed one second and then took the beaker out. Anya measured the volume of water, then tipped it away and did this twice more. Anya's results are shown in Table 10.6.

❶ What is the mean (average) volume of water collected?

To work out the **mean**:
- add all the results together;
- divide this number by how many results there are.

❷ Why was there a big difference between the volumes of water collected?

Juspal said it would be better if Anya collected the water for 10 seconds. Dividing this by 10 would then tell her how much water was flowing in one second.

❸ Would Juspal's method have made Anya's investigation more accurate or more precise?

Anya continued her investigation by:
- turning the tap to give a steady flow of water;
- writing down the voltmeter reading;
- collecting the water for 10 seconds;
- increasing the water flow;
- taking the two measurements again.

The results of Anya's investigation are given in Table 10.7 and drawn as a graph in Figure 10.28.

Voltage	Volume of water collected in 10 s	Volume of water collected in 1 s
0.25 V	680 ml	68 ml
0.30 V	720 ml	72 ml
0.50 V	780 ml	78 ml
0.60 V	880 ml	88 ml
0.70 V	940 ml	94 ml

Table 10.7 Anya's final results

❹ Anya only took one set of results. Why do you think it would have been hard to repeat the experiment and obtain a similar set of results?

❺ What did Anya do to make sure her investigation was a fair test?

❻ What happens to the voltage when the volume of water hitting the turbine each second increases?

❼ What volume of water do you think would be needed to generate 0.80 volts? To find out, continue the line in the graph up to 0.80 volts. This is called **extrapolating** the results.

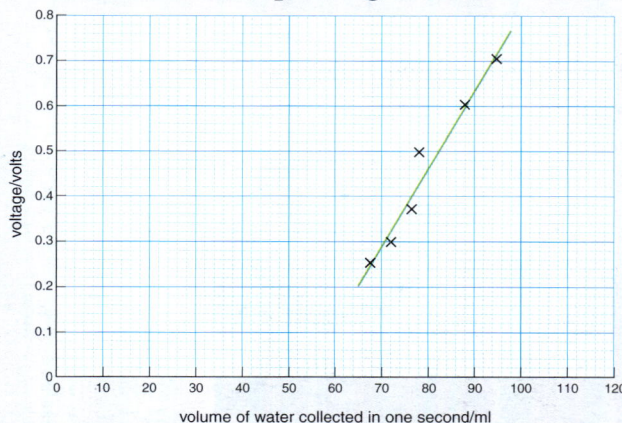

Figure 10.28 The results from Anya's investigation are shown in this graph.

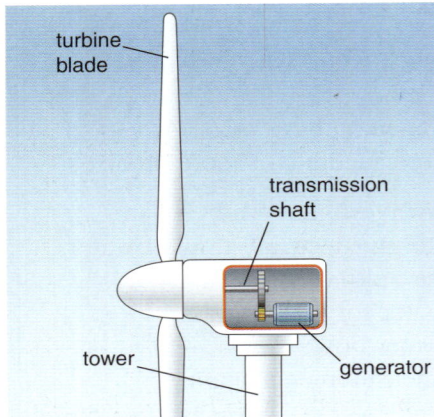

Figure 10.29 Inside a wind-powered generator

Figure 10.30 Using a wind generator to charge the batteries of a boat

A **wind farm** is a large number of wind generators grouped together.

Figure 10.32 Windy places are often in places of natural beauty.

Using the wind

Energy from the wind is not a new idea. More than a thousand years ago, windmills were used to grind grain.

A modern wind generator transforms the kinetic (movement) energy of the wind into electrical energy. As the wind blows, the turbine blades rotate. The turbine turns the generator which produces electricity (Figure 10.29).

One problem with wind generators is that no electricity is produced when the wind is not blowing. This makes the wind an unreliable energy resource.

Wind generators come in different sizes. Small units generate only enough to charge a battery. Large **wind farms** provide power to the National Grid. Only about 1% of Britain's electricity is generated using the wind, but it could increase to about 10%.

places where there is usually a strong wind

Figure 10.31 Wind turbines are built in places where there are strong winds. This makes generating electricity more reliable.

Wind farms need to be where it is windy! This is usually at the top of hills or at the coast. Some people think that wind farms are ugly and spoil the view. These people think wind farms cause visual pollution.

Figure 10.33 An offshore wind farm

Wind generators are not silent. Noise from the machinery and the rotating turbine blades is quite loud. Some people think wind farms cause noise pollution.

If turbines were built offshore, the wind could be used to generate more of our electricity. So far, three areas have been approved as sites for offshore wind farms. One of these, at Scroby Sands off the Norfolk coast, will use 39 giant turbines to generate electricity for 50 000 homes.

Activity – For or against wind farms

Local opposition has stopped a wind farm being built on coastal marshland near the Essex villages of Bradwell-on-Sea and Tillingham.

People living in the villages and nearby farms said the wind farm would:
- destroy the coastal marshes;
- threaten thousands of migrating birds;
- create noise pollution from the turbines.

❶ Imagine you are a scientist working for a wind power company. What could you say to persuade people to accept a wind farm in their area?

❷ Do scientists only produce results that support the company they work for? Answer 'yes' or 'no' and explain your answer.

❸ Do environmental groups only look for evidence to support their own views? Give reasons for your answer.

㉒ Which is a more reliable way to generate electricity, using tides or the wind? Give a reason for your answer.

㉓ Copy and complete the flow diagram in Figure 10.34 showing the energy changes at a pumped storage power station. Use the following words to complete the blanks:

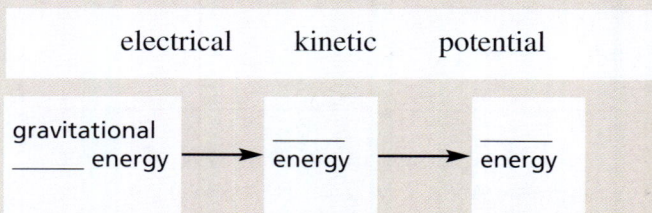

electrical   kinetic   potential

gravitational _____ energy → _____ energy → _____ energy

Figure 10.34 Energy changes at a pumped storage power station

㉔ The inhabitants of a small island need to replace their old power station. Half the people want to build a coal-burning power station. The other half want a wind farm.

Copy and complete Table 10.8 giving reasons for and against each suggestion. Some reasons have been filled in for you.

	For	**Against**
Coal-burning power station		It pollutes the air
Wind farm	It is renewable energy	

Table 10.8

Figure 10.35 Solar cells produce the electricity to run this satellite.

Figure 10.37 A solar cell provides the small amount of electricity a calculator needs.

10.7 Energy from the Sun and Earth

Energy direct from the Sun

Solar cells (sometimes called photocells) produce electricity directly from the Sun. But, if it's dark or cloudy, very little electricity is produced. So solar cells are an unreliable way to produce electricity.

Figure 10.36 A solar-powered MP3 player

Sometimes, solar cells are the best way of producing electricity. They are used in remote places, on satellites or in devices where only small amounts of electricity are needed.

---

❷❺ Why are solar cells suitable for use with:
   a) a mobile phone charger?
   b) a satellite?

❷❻ In the UK, one square metre (1 m²) of solar cells can generate 40 W of power.
   a) How much land has to be covered by solar cells to generate 1 kW of power? (Remember, 1 kW = 1000 W.)
   b) A medium-sized power station generates 2000 MW. How much land has to be covered by

solar cells to generate 2000 MW of power? (Remember, 1 MW = 1000 kW.)
   c) Would it be possible to build a solar power station in the UK? Give a reason for your answer.

❷❼ An African village uses solar cells to generate the electricity needed to operate the pump at a water well. Why are solar cells used to generate the electricity rather than a petrol generator?

Figure 10.38 Letting off steam the natural way!

28 Copy and complete the sentences below using the words in this box.

| efficient | electrical | light | wasted |

The energy input to a solar cell is _____ energy. The useful output from the solar cell is _____ energy. A solar cell is about 20% _____, this means 80% of the input energy is _____.

29 Copy the energy resource statements given below. Then copy the methods of producing electricity. Draw an arrow connecting each resource with the correct method of producing electricity.

Energy resource statements

Produces a lot of noise

Land may be flooded

Built across a river estuary

Methods of producing electricity

Tidal barrage

Wind generator

Falling water (hydroelectric)

Energy from the Earth

Radioactive elements like uranium are found inside the Earth. When atoms of these radioactive elements decay (see Sections 11.8 and 11.9) they transfer heat to the surrounding rocks. This heat in the underground rocks is called **geothermal energy**.

In some places where there are volcanoes, geysers or hot springs, water heated by geothermal energy often reaches the Earth's surface as steam. This steam can be used to drive turbines, connected to electricity generators.

30 Copy and complete the following sentences.

Geothermal energy comes from hot underground _____. The heat comes from the decay of atoms in _____ elements. Just like a thermal power station _____ is used to drive a turbine.

31 The graph in Figure 10.39 shows how the demand for electricity in the UK changes over 24 hours.
a) Why is more electricity needed after 6.00a.m.?
b) What type of power station can be used to supply the extra power needed at peak times?
c) Why do electricity generating companies need to know when popular TV programmes start and finish?

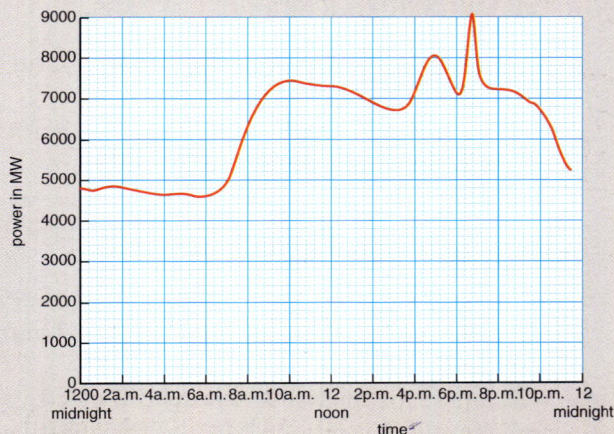

Figure 10.39

Summary

✓ Electrical energy is easily transformed into other forms of energy.

✓ **Power** is the rate at which a device transforms energy.

✓ The power of an **appliance** is measured in **watts** (W) or **kilowatts** (kW).

✓ 1 kilowatt (kW) = 1000 watts (W).

✓ 1 watt (W) = 1 joule / second (J/s).

✓ The amount of electrical energy that an appliance transforms depends on the power of the appliance and how long it is used.

✓ The amount of electricity transferred from the mains can be calculated using the equation:

energy transferred = power × time
(in kilowatt-hours) (in kilowatts) (in hours)

✓ The **National Grid** is a network of cables and transformers that link power stations to consumers.

✓ **Non-renewable** and **renewable energy sources** can be used to generate electricity.

✓ The commonest energy sources are **fossil fuels** – coal, oil and natural gas.

✓ Fossil fuels are non-renewable energy sources.

✓ Renewable energy sources include the wind, tides, waves, falling water in hydroelectric schemes, solar energy from the Sun and geothermal energy from hot underground rocks.

✓ In thermal power stations an energy source is used to heat water. The steam produced drives a turbine, which turns an electrical generator.

✓ The energy from renewable sources can be used to drive turbines directly.

✓ Each type of energy source has its advantages and disadvantages.

EXAMQUESTIONS

❶ The following sentences describe how electricity is generated and sent from a power station to a home. The sentences are in the wrong order.
Copy the sentences, putting them in the right order. Start with sentence E.
  A A transformer steps-up the voltage to 400 000 volts.
  B Energy is sent through the National Grid at this very high voltage.
  C The voltage is reduced to 240 volts before the electricity goes to our homes.
  D Steam drives a turbine.
  E Coal is burnt to heat water and produce steam.
  F A turbine turns a generator and electricity is made.   (4 marks)

❷ William keeps his house warm with electrical night storage heaters. In March 2005, William had his loft insulated. Since then the amount of electrical energy he uses has gone down. Table 10.9 gives the meter readings taken from three of his electricity bills.

Date of bill	Meter reading (kW h)
March 2004	67 000
March 2005	68 850
March 2006	70 000

Table 10.9

a) How many kilowatt-hours of energy did William use between March 2004 and March 2005?   (1 mark)
b) How many kilowatt-hours of energy did William use between March 2005 and March 2006?   (1 mark)
c) How much money did the loft insulation save William in one year between March 2005 and March 2006? (One kilowatt-hour of energy costs 12p.)   (2 marks)

d) The loft insulation cost William £240. How many years will it be before William has paid for the insulation from his energy savings? *(2 marks)*

❸ Figure 10.40 shows the percentage of Britain's electricity generated by different energy resources.

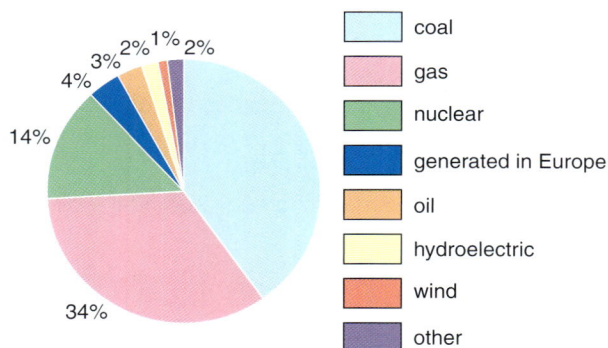

Figure 10.40 Britain's electricity is generated using a range of energy resources.

a) i) What percentage of Britain's electricity is generated in coal power stations? *(1 mark)*

ii) Name two ways of generating electricity that do not involve burning a fuel. *(2 marks)*

iii) Name one energy resource used to generate electricity that is not given in Figure 10.40. *(1 mark)*

b) Give one advantage and one disadvantage of generating electricity using nuclear fuel rather than coal. *(2 marks)*

❹ Figure 10.41 shows an advert for solid fuel firelighters.

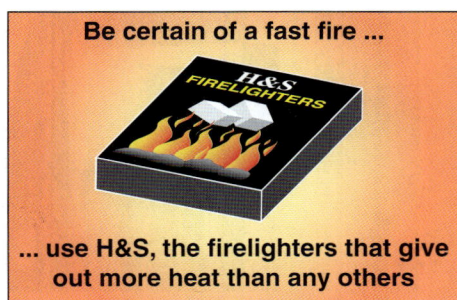

Be certain of a fast fire ...

... use H&S, the firelighters that give out more heat than any others

Figure 10.41

a) Paul decides to test the advert. He plans an experiment to compare the heat given out by H&S firelighters with two other brands.

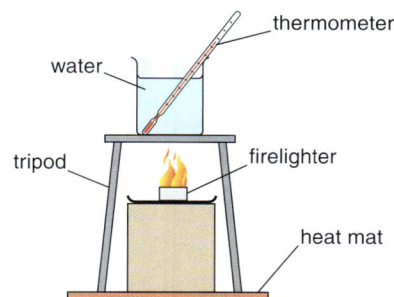

Figure 10.42 The apparatus used in Paul's experiment

Here is Paul's plan for the experiment.

- Take 1 g of H&S firelighter and place it on a tin lid.
- Put 80 ml of water into a beaker.
- Measure the temperature of the water.
- Use a match to set fire to the firelighter and then use the burning firelighter to heat the water.
- When all the firelighter has burnt, measure the new water temperature.
- Repeat the experiment with the other two brands of firelighter.

i) Is the type of firelighter a categoric or continuous variable? *(1 mark)*

ii) Name two controlled variables. *(2 marks)*

iii) Suggest one change Paul could make to his choice of measuring instruments to improve the accuracy or precision of the experiment. *(1 mark)*

b) To compare his results, Paul drew the bar chart shown in Figure 10.43.

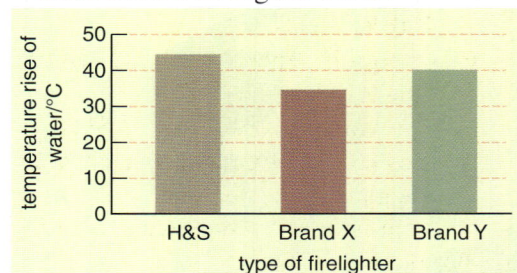

Figure 10.43 A bar chart showing the results from Paul's experiment

Are Paul's results good enough for him to confirm that H&S firelighters give out more heat than any others? Give a reason for your answer. *(2 marks)*

Chapter 11
What are the properties, uses and hazards of electromagnetic waves and radioactive substances?

At the end of this chapter you should:

- ✓ know that electromagnetic waves transfer (move) energy from one place to another;
- ✓ understand that electromagnetic waves and radioactive substances have many uses;
- ✓ know that the use of electromagnetic and nuclear radiations involve hazards as well as benefits;
- ✓ know about the ways of reducing exposure to different types of electromagnetic and nuclear radiation;
- ✓ know that atoms have a small central nucleus containing protons and neutrons surrounded by electrons;

- ✓ know that radioactive substances give out three types of radiation (alpha particles, beta particles and gamma rays) from the nuclei of their atoms;
- ✓ understand the important properties of alpha particles, beta particles and gamma rays;
- ✓ understand the term 'half-life';
- ✓ know that communication signals can be analogue or digital.

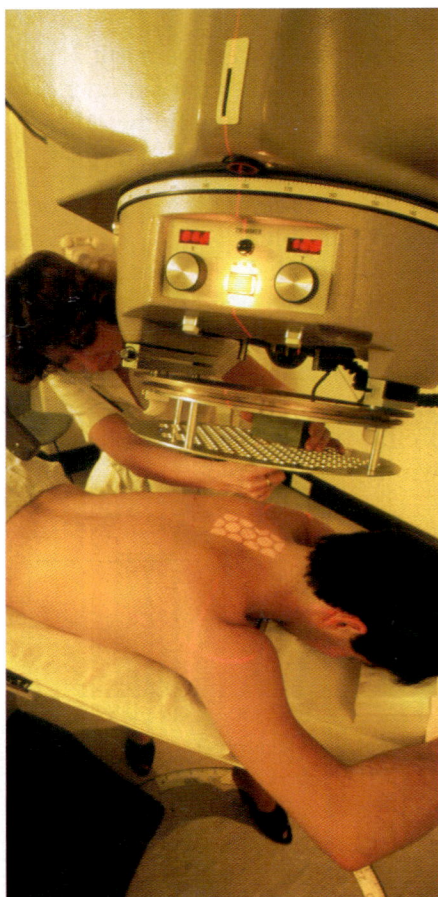

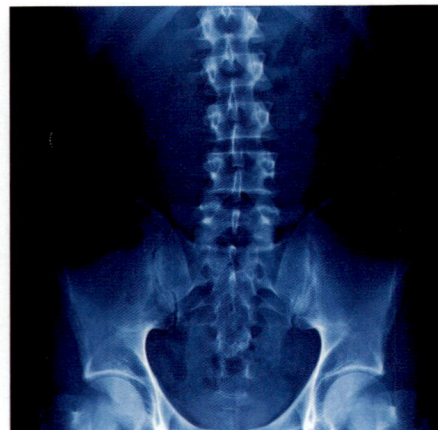

Figure 11.1 Electromagnetic waves and radioactive substances are very useful but they may also be very dangerous.

11.1 Looking at waves

In Figure 11.3 John and Sarah have stretched a 'slinky' spring across the floor. John holds his end of the slinky still. Sarah moves her end of the slinky from side to side. This sends a series of waves (pulses) down the slinky. Each pulse carries energy, but the slinky itself does not move from Sarah to John. This simple experiment shows that:

waves can transfer energy from one place to another without moving material from one place to another.

This is just like the Mexican wave at a sports event. As the Mexican wave travels around the stadium it looks like something is moving, but the spectators stay where they are.

Figure 11.2 A Mexican wave travels around the stadium.

energy transfer

movements of hand from side to side

Figure 11.3 Waves (pulses) moving down a slinky

Raindrops falling into a pond make ripples on the water (Figure 11.4). It looks as though something is moving along, but it's not the water. Energy is moving outwards with the ripples, but not the water. The water just moves up and down. It does *not* move to the edge of the pond. A cork on the surface of the pond just bobs up and down with the water.

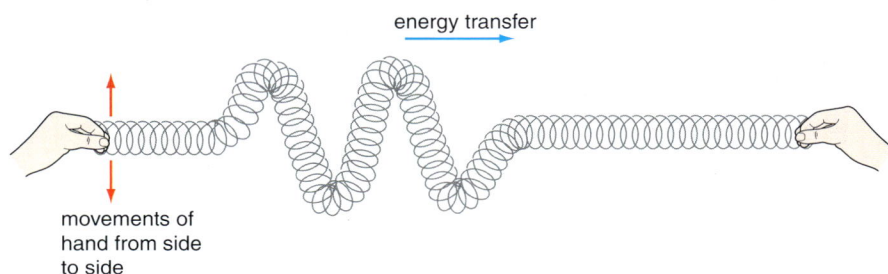

Figure 11.4 Raindrops cause ripples to spread across the pond's surface.

Waves can be described by three quantities – **wavelength**, **frequency** and **amplitude**. Looking at a wave makes it easier to understand what these quantities are (Figure 11.5).

Frequency is measured in **hertz** (Hz). A source making one wave every second has a frequency of 1 hertz (1 Hz). If there are 50 waves in 10 seconds, the frequency is 5 waves per second or 5 Hz.

Wavelength is the distance from a point on one wave to the same point on the next wave.

Frequency is the number of waves that are made each second. It is also the number of waves passing a point each second.

Amplitude is the distance from the middle of a wave to the top or bottom of the wave.

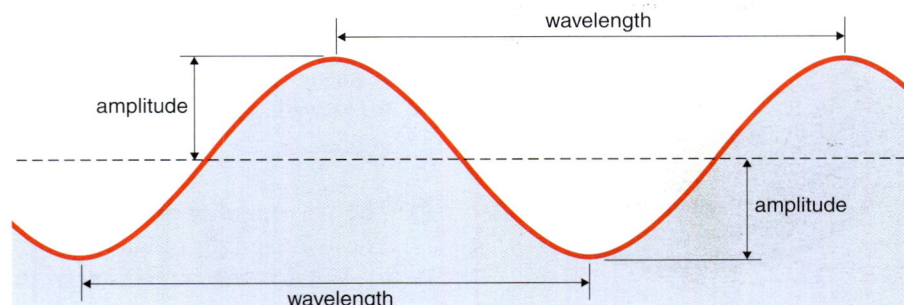

wavelength

amplitude

amplitude

wavelength

Figure 11.5 A wave moving across a water surface

A wave with a big amplitude moves more energy than a wave with a smaller amplitude.

Figure 11.6 The energy of waves at sea or from earthquakes and tsunamis can cause massive damage.

Wave speed and the wave equation

The wave speed is how far a wave moves in one second. If you know the frequency and wavelength of a wave, its wave speed can be worked out using the equation:

$$\text{wave speed} = \text{frequency} \times \text{wavelength}$$

wave speed in metre/second (m/s) frequency in hertz (Hz) wavelength in metre (m)

This **wave equation** works for all waves.

Example
Water waves with a wavelength of 0.6 metres make a boat bob up and down 2 times a second. At what speed do the waves move across the surface of the water?

$$\text{wave speed} = \text{frequency} \times \text{wavelength}$$
$$\text{wave speed} = 2 \times 0.6 = 1.2 \text{ m/s}$$

---

❶ Copy and complete the sentences below using the words in this box.

> energy   frequency   hertz

a) The _____ of a wave equals the number of waves going past a point every second.
b) A wave transfers _____ from one place to another.
c) The wave maker in a leisure pool makes 1 wave every 4 seconds. The frequency of the wave maker is 0.25 _____.

❷ Lucy is playing in her paddling pool. As she pushes a ball up and down in the water it makes a small wave.
a) Which way does the water move as the wave moves across the surface of the pool?
b) Why do the other balls in the pool bob up and down, but not move sideways?

Figure 11.7

❸ A stone thrown into a pond produces water waves of frequency 6 Hz with a wavelength of 5 cm. Calculate the speed of these waves.

❹ The wave maker at a leisure pool makes 1 wave every 2 seconds. The wavelength of each wave is 4 metres.
a) What is the frequency of the waves?
b) Calculate the speed of the waves across the surface of the water.

11.2 The electromagnetic 'family'

You already know a lot about one part of the **electromagnetic spectrum** – the part we call visible light. However, visible light is only a small part of the much larger electromagnetic 'family'. The electromagnetic family is a whole 'family' of waves that make up the electromagnetic spectrum.

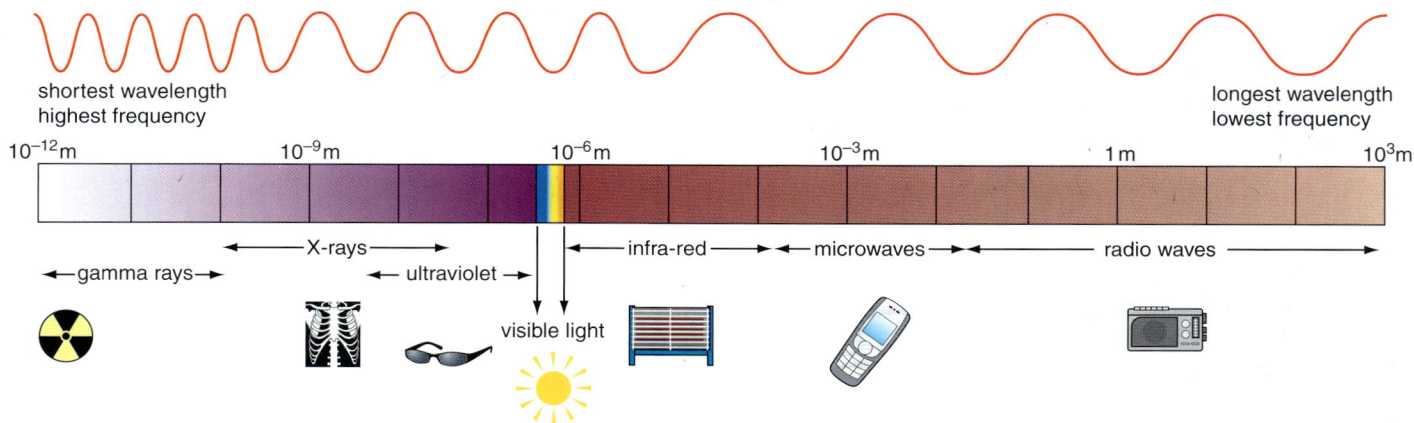

shortest wavelength
highest frequency

longest wavelength
lowest frequency

10^{-12}m 10^{-9}m 10^{-6}m 10^{-3}m   1 m 10^3m

gamma rays — X-rays — ultraviolet — visible light — infra-red — microwaves — radio waves

Figure 11.8 The electromagnetic spectrum – a 'family' of waves

5 A navigation system uses radio waves. The waves have a frequency of 1.5 MHz and a wavelength of 200 m. Use the equation below to show that the radio waves have the same speed as light (300 000 000 m/s). (1 MHz = 1 megahertz = 1 000 000 Hz.)

$$\frac{\text{wave}}{\text{speed}} = \text{frequency} \times \text{wavelength}$$

There are no gaps in wavelength in the electromagnetic spectrum. Wavelengths run smoothly from one value to another. This is a continuous spectrum. But notice how the wavelengths get longer as you go from left to right.

All electromagnetic waves:
- obey the wave equation (wave speed = wavelength × frequency);
- can travel through a vacuum (empty space);
- travel at the same speed (300 000 000 metres per second) through a vacuum or air – this means that waves with the longest wavelength have the lowest frequency and waves with the shortest wavelength have the highest frequency;
- transfer energy from one place to another;
- can be reflected, absorbed or transmitted.

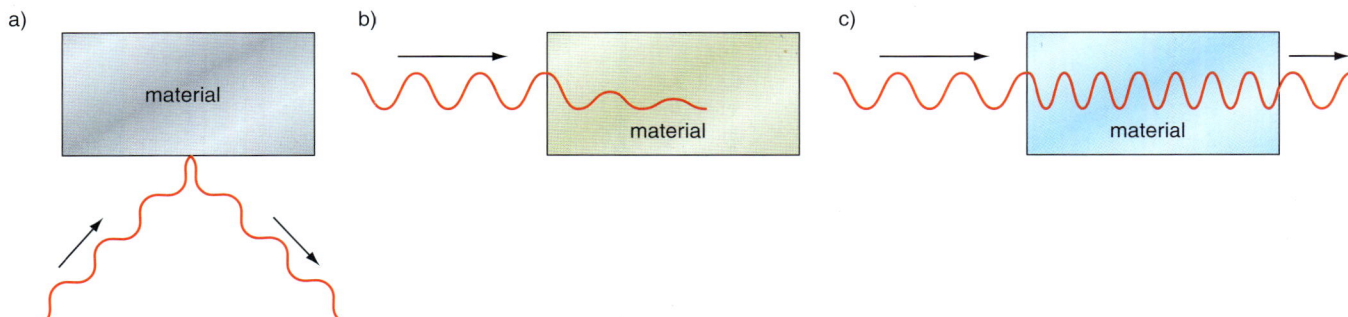

a) material

b) material

c) material

Figure 11.9 a) The waves are reflected by the material. b) The waves go into the material but do not come out. The waves have been absorbed c) The waves go into and come out of the material. The waves have been transmitted.

Different types of electromagnetic wave are reflected, absorbed or transmitted in different amounts by different materials.

Electromagnetic waves transfer (move) energy from one place to another. So, when electromagnetic waves are absorbed, the energy is given to the absorbing material. This may make the absorbing material hotter. It may even produce an alternating current (a.c.) of electricity in the absorbing material, with the same frequency as the radiation.

> **6** Which parts of the electromagnetic spectrum have a higher frequency than ultraviolet?

Figure 11.10 The energy carried by infra-red radiation soon warms us up.

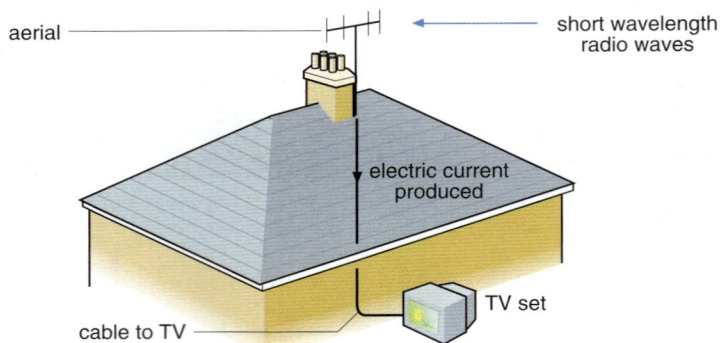

Figure 11.11 Electromagnetic radiation can produce an alternating current.

11.3 Using microwaves, infra-red and ultraviolet waves in the home

Microwave cooking

How does a microwave oven cook food? The answer might seem simple. You put the food in, push a few buttons and the food gets hot. But of course it's not that simple.

Microwaves pass easily through some materials such as plastic, paper and glass but they are absorbed by water molecules. So the energy carried by microwaves can heat water up. As most foods contain a lot of water, the microwaves heat this, which then cooks the food.

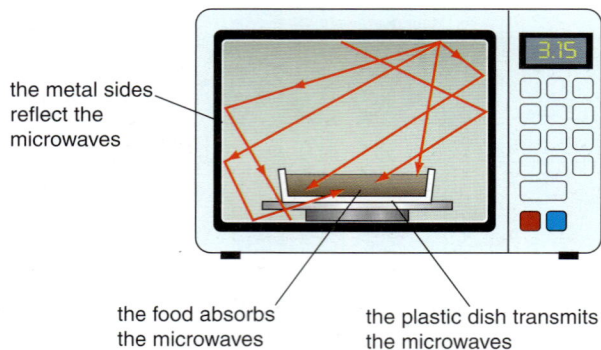

Figure 11.12 The microwaves are reflected by the walls of the oven, transmitted through the plastic dish and absorbed by the food.

Food cooked in a gas or electric oven is heated by conduction. This is quite a slow process. Microwaves heat the food more quickly by penetrating several centimetres into the food before being absorbed.

---

7 Copy and complete Table 11.1 by writing each of the following objects in the correct column. The first two have been done for you.
paper plate, tin can, aluminium foil, glass dish, pottery mug, steel knife, plastic bowl

Objects that transmit microwaves	Objects that reflect microwaves
Paper plate	Tin can

Table 11.1

8 Copy and complete the following sentence using one of the statements A, B or C.

Microwave ovens waste less energy than electric ovens because
A   energy is not used to heat up the inside of a microwave oven.
B   microwave ovens are usually smaller than electric ovens.
C   electric ovens have a higher power than microwave ovens.

9 Why is it important that microwaves don't leak from a microwave oven?

Activity – Sharon investigates microwaves

Sharon noticed that all the food containers she uses in her microwave are made from plastic. She thinks this is because microwaves lose less energy going through a plastic container than they do through a glass container.

To test her idea, Sharon poured 250 cm³ of water into a plastic beaker. She put the beaker into a microwave oven and heated it for 2 minutes. Every 20 seconds Sharon stopped the microwave and took out the beaker. She stirred the water and took its temperature. Then Sharon did the experiment again. This time she heated 250 cm³ of water in a glass beaker.

Table 11.2 shows Sharon's results with the plastic beaker.

Time in seconds	0	20	40	60	80	100	120
Temperature in °C	22	30	40	47	52	66	74

Table 11.2 Sharon's results with the plastic beaker

1 Use Sharon's results to draw a graph of temperature (vertically) against time (horizontally). When you have plotted all the points draw a line of best fit. (Remember a line of best fit does not have to go through all the points.)

2 One point on the graph does not fit the pattern. Draw a circle around this point.

3 Unscramble the letters AALMNOOSU to give the word used to describe a result that does not fit the pattern.

4 Copy and complete the following sentence. Sharon stirred the water to make sure it was all at the _____ temperature.

5 What was the temperature of the water after 50 seconds?

6 Use your graph to estimate how long it would have taken the water to boil.

7 In Sharon's investigation, which quantity was the **dependent variable**: time or temperature?

8 Write down one thing that Sharon did to make her investigation a fair test.

9 The thermometer Sharon used could measure to the nearest 1 °C. If Sharon had used a thermometer that could measure to the nearest 0.1 °C, would her measurements have been *more accurate, more precise* or *more reliable*?

10 If Sharon is right, will the temperature of the water in the glass beaker be higher or lower than 74°C after 120 seconds?

Figure 11.13 The infra-red radiation emitted by the heating element is absorbed by the bread.

Infra-red radiation for cooking and detection

All objects emit (give out) and absorb (take in) infra-red radiation. The hotter the object, the more infra-red it emits. Objects that absorb infra-red get hotter.

We cannot see infra-red radiation. But an infra-red camera can be used to 'see' warm objects even when we can't see them in the dark. This type of camera is used by the Army and Police to spot people in the dark. It is also used by fire crews to find people trapped in smoke-filled buildings.

Ultraviolet (UV) radiation – sunbathing and energy-efficient lamps

Any object with a high enough temperature emits UV radiation. The Sun emits UV radiation. UV radiation passes through some substances but is absorbed by others. Most sorts of glass are very good absorbers of UV radiation.

Figure 11.14 Being ready to ski means wearing the right eye protection!

Some surfaces are good reflectors of UV radiation. If you go skiing you need to remember that snow reflects up to 90% of the UV that hits it. As UV can damage your eyes, you need to wear glasses or goggles that give the right protection. Without these, you could end up with 'snow blindness'.

Some of the UV given out by the Sun is absorbed by ozone in the Earth's atmosphere. The UV that does get through gives you a suntan. You get a tan because one type of skin cell produces a substance called melanin. Melanin is dark brown. It makes your skin change colour. By absorbing UV, melanin helps to protect other body cells from damage. But beware – melanin is not produced instantly. It takes time.

Absorbing too much UV from the Sun (or a sunbed) is dangerous. It increases your risk of getting skin cancer and may damage your eyes. So what should you do to limit your exposure to UV? Here are a few suggestions:

- listen to weather reports that give the UV level (the 'sun-index');
- wear a hat with a wide brim;
- wear Sun-protective clothing;
- stay out of the Sun between 10a.m. and 3p.m;
- slap on lots of high SPF (Sun Protection Factor) sunscreen cream.

It's not all bad news. UV lets your body produce vitamin D, a vitamin essential for healthy growth.

Figure 11.15 Sunbathing can be very pleasant, but you must take care not to damage your skin.

Energy-efficient lamps

Some substances absorb ultraviolet radiation and then emit (give out) the energy as visible light. This is called **fluorescence**. Fluorescent paints, dyes and security markers all work like this. They seem to glow.

Fluorescent lamps work by producing ultraviolet radiation. The chemical covering the inside of the glass absorbs the ultraviolet radiation. It then gives out this energy as visible light and appears to glow.

Figure 11.16 The fluorescent dial of this watch glows in the dark.

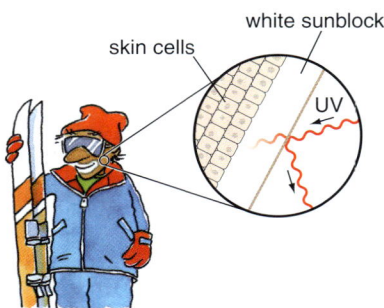

Figure 11.17 The sunscreen cream and sun block helps protect George from skin cancer.

Figure 11.18 Why can't you get a suntan inside a glass conservatory?

⑩ Copy and complete the sentences a, b, c, d by choosing the correct ending from w, x, y and z.
   a) Infra-red radiation is used . . .   w) glow in the dark.
   b) Things that absorb infra-red . . .   x) get hotter.
   c) Fluorescent chemicals . . .   y) skin cancer.
   d) UV radiation can give you a suntan   z) to toast bread.
      but also . . .

⑪ Why are fluorescent lamps often described as 'energy-efficient'?

⑫ Before skiing, George puts sunscreen cream on his face. He also puts white sun block on his lips. Figure 11.17 shows how the sun block helps to protect George from harmful UV rays.
   a) Explain how the sun block helps to protect George from the UV rays.
   b) Why is it important that George puts a thick layer of sunscreen cream onto his face?
   c) Why does George need the sunscreen cream even though it is not hot where he is skiing?

Activity – Sensible sunbathing

A recent report said that travel agents are not doing enough to warn their customers about the dangers of sunbathing. Although the number of deaths from skin cancer is going up, travel brochures still show lots of sun-tanned people lazing around in the Sun.

Write a short article or design a poster for a travel brochure. The article or poster must warn people about the risk of skin cancer and must give advice on how to reduce this risk. Of course, the article or poster must not scare people so much that they stop wanting to have a holiday in the Sun!

11.4 Using X-rays and gamma rays in medicine

X-rays and gamma rays are electromagnetic waves with a short wavelength. They are very penetrating. Both types of radiation can pass through some substances with very little energy being absorbed. Just like visible light, they also make photographic film go black.

When X-rays and gamma rays are absorbed, their energy can remove electrons from some atoms. This is called **ionisation**. So both X-rays and gamma rays are **ionising** radiations. This makes them harmful to us. Too much exposure to X-rays or gamma rays may damage our central nervous system, cause genes to mutate, or even cause cancer. So we must be exposed to X-rays and gamma rays as little as possible.

X-ray photography

X-rays easily pass through healthy skin and flesh. But bones, teeth and diseased tissues absorb X-rays. In an X-ray photograph any bones, teeth or diseased tissue will stand out because they stop the X-rays turning the photographic film black. This makes X-rays very useful. Doctors use them to show where a bone is broken, and dentists use them to show tooth decay.

Radiographers are the people who take X-ray photographs in hospitals. They must be protected from the dangerous effects of X-rays. Often they work behind lead or thick glass screens. These materials are very good at absorbing X-rays. If they cannot work behind a protective screen, then they wear a lead-lined apron. Lead is also used to protect the parts of a patient's body that are not being X-rayed.

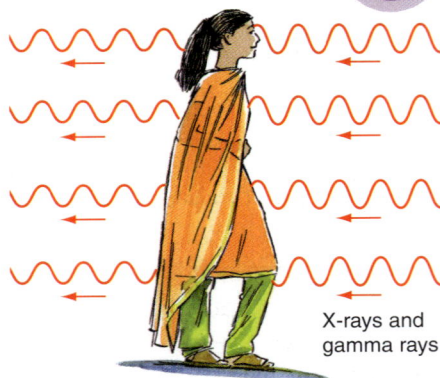

Figure 11.19 X-rays and gamma rays can go straight through you. But some energy will be absorbed by your body cells. This makes X-rays and gamma rays dangerous.

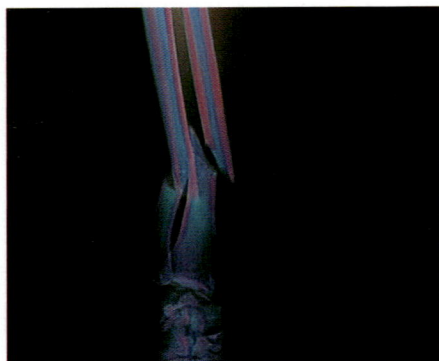

Figure 11.20 An X-ray photograph of a broken arm

⑬ Copy and complete the following sentences.
  a) X-rays will pass _____ skin and _____ but are absorbed by bones and _____ tissue.
  b) X-rays affect _____ film, making it go _____.
  c) Radiographers are protected from X-rays by _____ or thick glass. These materials _____ X-rays.

⑭ What precautions should a dentist take before X-raying a patient's tooth?

Gamma rays

Gamma rays are used to diagnose and treat diseases such as cancer.

To diagnose cancer, a small amount of radioactive material is injected into the patient's body. The gamma rays given out go through the patient's body and are detected by a gamma camera.

Figure 11.21 Exposure to X-rays should always be kept to a minimum.

Gamma rays are used to treat cancer because in large doses they penetrate and kill cancer cells.

The properties and uses of gamma rays are studied further in Sections 11.9 and 11.12.

15 Which of the following statements are true and which are false? Copy out the true statements. Rewrite the false statements to make them true.
   A Gamma rays have a higher frequency than X-rays.
   B X-rays have a shorter wavelength than gamma rays.
   C Gamma rays travel faster through the air than X-rays.
   D X-rays will pass easily through a gold ring.

16 Explain why gamma rays from outside the body can kill cancer cells inside the body.

Radiographer

What does a radiographer do? Radiographers diagnose and treat illnesses. During diagnosis, radiographers use different techniques to produce images of the body. Using X-ray machines is just one part of the job. During the treatment of illnesses, radiographers will be involved in planning and giving treatment using high-energy X-rays and gamma rays. Working as part of a team, radiographers need good communication skills. If you want to find out more about a career as a radiographer go to www.radiographycareers.co.uk or www.sor.org

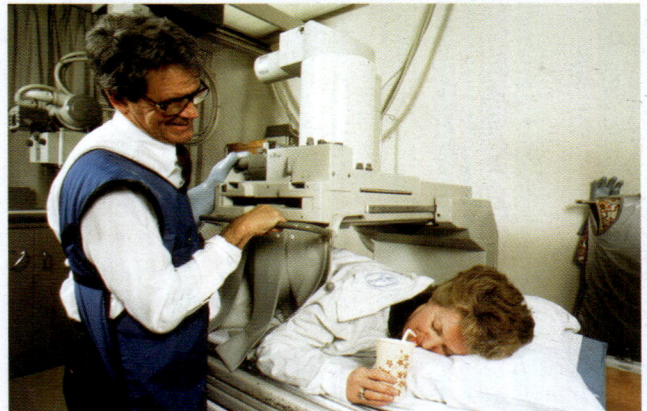

Figure 11.22 This radiographer is wearing a protective lead-lined apron. He is preparing a patient for an X-ray.

11.5 Using waves in communications

Visible light, infra-red, microwaves and radio waves are all used for communications.

Visible light and infra-red

Optical fibres are usually made of glass. They are very thin and flexible. The fibres can carry infra-red and visible light rays.

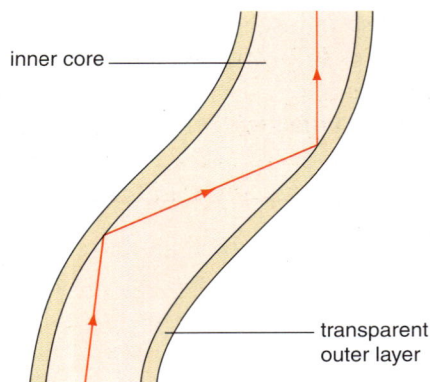

inner core

transparent outer layer

Figure 11.23 Reflection allows visible light and infra-red to pass along the optical fibre.

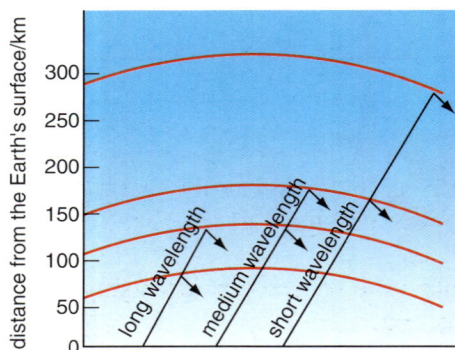

Figure 11.24 Radio waves with a longer wavelength than 10 metres are reflected by the ionosphere.

Figure 11.25 Stereo radio and TV programmes are transmitted from telecommunications towers, such as the BT Tower in London.

Infra-red or visible light rays going in at one end of the fibre are reflected many times until they come out at the other end of the fibre. The rays can even follow any bends and twists in the fibre.

For communications, digital signals are produced by changing speech and electronic messages into light or infra-red signals (see Section 11.6). These signals can then be sent through the optical fibres. Many telephone systems now use optical fibres rather than copper cables. A single fibre can carry thousands of telephone conversations all at the same time.

Microwaves and radio waves

The Earth's atmosphere has an electrically charged layer (the **ionosphere**). This layer can reflect radio waves that have a longer wavelength than 10 metres. By using this reflection, radio waves can be sent around the world, even though the Earth's surface is curved.

Radio and television programmes that are broadcast in stereo use radio waves with a very short wavelength and a very high frequency. The ionosphere does not reflect these waves. The waves go straight from one transmission station to another. At each station, the signal is transmitted in all directions to our homes.

Microwaves with very short wavelengths (only a few centimetres) pass through the Earth's atmosphere. So information is carried by microwaves to and from satellites. Many people now receive television programmes from signals carried by microwaves. These have travelled from a transmitter on Earth to a satellite above the Earth and then back to a dish on the side of the person's house.

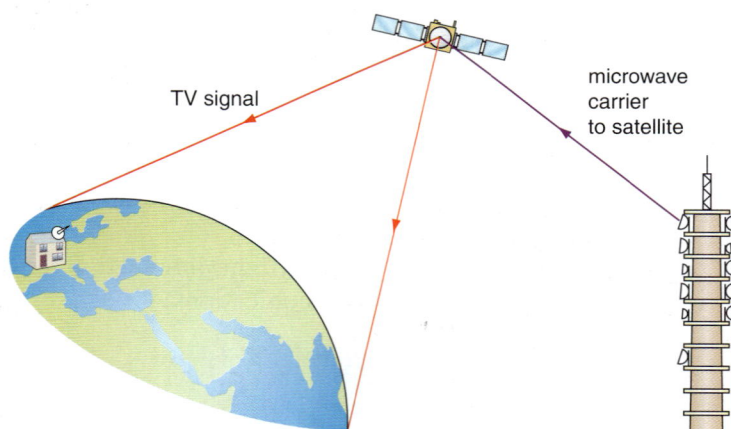

TV signal

microwave carrier to satellite

Figure 11.26 A satellite can transmit a microwave signal that covers a large part of the Earth's surface.

Nowadays, the most common use of microwaves is in mobile phone networks. These microwaves are transmitted from place to place via tall aerial masts or from one continent to another via satellites.

17 Draw a diagram to show the path of a pulse of infra-red radiation through an optical fibre.

18 A TV programme in the UK uses a live satellite link to question a reporter in the USA. Why is there a delay between the question being asked and the reporter hearing it?

19 Copy and complete the following sentence by choosing the correct ending.
Microwaves are able to transmit information to a satellite because they *are absorbed by water molecules / pass through the atmosphere / travel at the same speed as light.*

Activity – Are mobile phones a health risk?

Mobile phones send and pick up messages using microwaves. Some of the microwaves move towards us and go into our bodies, including our heads. As we learnt in Section 11.3, microwaves are absorbed by water, and our bodies contain a lot of water. The energy absorbed by the water will raise the temperature of our body tissues and this may damage our body cells. This is why using a mobile phone could be a health risk. But no one knows for certain.

Figure 11.27 Some energy from the microwaves is absorbed by the head and body.

Some research has suggested that there could be a link between using a mobile phone and health problems such as headaches, memory loss and brain tumours. But, at the moment, there is no reliable evidence that using a mobile is definitely harmful. Studies show that more research is needed.

In a report produced in 2005, Professor Sir William Stewart of the National Radiological Protection Board said

'I don't think we can put our hands on our hearts and say mobile phones are safe. If there are risks –

and we think there may be risks – the people who are going to be most affected are children, and the younger the child, the greater the danger.'

Following this report, shops in the UK stopped selling a mobile phone designed for young children. Some people did not like this decision. Many parents want their children to have a mobile phone for emergencies. They say that they worry less if their children can call them at any time.

1 Copy and complete the sentence below using the words in this box.

reliable	same	scientists

When scientist produce _____ evidence it means that other _____ doing exactly the same experiment would get exactly the _____ results.

2 What health problems may be caused by using a mobile phone?

3 What reason might a manufacturer give for making a mobile phone for young children?

4 Copy and complete the following sentence by choosing the correct ending.
A young child is more likely to be affected by the radiation from a mobile phone than an adult because
they will talk longer on the phone.
they will send more text messages.
their skulls are thinner, so more energy gets through to their brain.

5 If scientific evidence showed that using a mobile phone increased the risk of getting a brain tumour, would you still use one? Give reasons for your answer.

11.6

Why are digital signals taking the place of analogue signals?

People have always wanted to communicate, and nowadays it is easier than ever. Just pick up the phone, use your mobile, or send an e-mail. We can communicate with each other anywhere in the world and it's almost instant. But it hasn't always been as easy or as fast as this.

The telegraph was the first modern invention that allowed information to be sent using electricity as a message carrier. The signal was sent using a series of coded pulses of electricity. This type of signal is called a **digital signal**. In modern terms the telegraph was very slow. At one end, an operator needed to code the message then send the signal. At the other end, another operator wrote down the signal then decoded it.

> A **digital signal** is a series of ON and OFF pulses.

In modern digital systems, electronic circuits have replaced the operators. These circuits produce signals that can only be 'ON' or 'OFF'. These are called the **states** of a digital signal. The number 1 is used to represent the 'ON' state. The number 0 is used to represent the 'OFF' state. This means any digital signal can be written as a sequence of 1s and 0s.

The telephone, invented in 1876, was the next big breakthrough in electrical communications.

1 0 1 1 0 1 0 1 1 0 1

Figure 11.28 A digital signal is either 'ON' or 'OFF'.

When you speak, you produce sound waves. A microphone inside the telephone turns the sound waves into electrical signals. The amplitude and frequency of the electrical signals change as the amplitude and frequency of the sound waves change. These signals are called **analogue signals**.

> **Analogue signals** are continuous waves that vary in amplitude and/or frequency.

In older telephone systems, copper cables carry the analogue signals between the person talking and the person listening. The earphone in the telephone is like a small loudspeaker. It changes the electrical signals back into sound. In modern telephone systems, optical fibres, microwaves and radio waves are replacing the copper cables linking 'sender' to 'receiver'. (See also Section 11.5.) If an optical fibre link is used, a digital signal is sent as a series of light pulses.

So communication signals may be either digital or analogue. But using digital signals has two main advantages over using analogue signals.

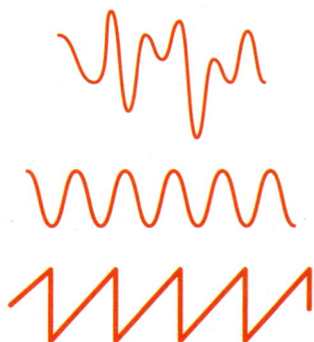

Figure 11.29 All these waves are analogue signals. The waves vary in amplitude and/or frequency.

- Information carried as a digital signal does not change during transmission. The signal that is received is the same as the signal that was sent out. This means the received signals are high quality. When an analogue signal is transmitted it loses quality.
- Computers work and communicate using digital signals. This makes it easy for a computer to process data in the form of a digital signal.

20 Figure 11.30 shows two signals. One is analogue and one is digital.
a) Copy the two signals. Label each one with its correct name.
b) What are the main differences between an analogue signal and a digital signal?
c) Give two advantages of sending a digital signal rather than an analogue signal.

Figure 11.30

21 Telephone companies are replacing underground copper cables with optical fibre cables.
a) What type of signal is sent along an optical fibre: analogue or digital?
b) Which two of the following are used to carry information along an optical fibre?

infra-red   microwaves   visible light
ultraviolet

c) Give one reason why telephone companies are changing from using copper cables to using optical fibre cables.

11.7 What are atoms really like?

Over a century ago, scientists thought that atoms were hard, solid particles like tiny marbles. Then, in 1897, J.J. Thomson discovered that atoms contained tiny negative particles, which he called **electrons**. As atoms were neutral overall, Thomson's discovery of negative electrons led scientists to think that some parts of an atom must be positive. In 1909, Ernest Rutherford found evidence for positive particles in the centre (the **nucleus**) of atoms. Rutherford called these positive particles **protons**.

Further experiments showed that the nuclei of atoms must contain neutral particles as well as protons. These neutral particles were discovered in 1932 and called **neutrons**.

We now know that:
- all atoms are made up from three particles – protons, neutrons and electrons;
- the nuclei of atoms contain protons and neutrons;
- protons and neutrons have the same mass – the proton and neutron are each given a relative mass of one;
- protons have a positive charge, but neutrons are neutral (they have no charge);
- more than 99% of an atom is empty space in which there are negative electrons;
- the mass of an electron is about 2000 times less than that of a proton;
- the negative charge on one electron just cancels the positive charge on one proton;
- electrons whiz around the nucleus very rapidly.

Figure 11.31 J.J. Thomson has been called 'the father of modern science'. Thomson discovered electrons in 1897 and this opened up the study of the structure of atoms. Thomson and his students at Cambridge University won eight Nobel Prizes for Physics or Chemistry.

Figure 11.32 Ernest Rutherford left New Zealand to work with J.J. Thomson at Cambridge University in 1895. Later, he moved to Manchester University where he won the Nobel Prize for Chemistry. Rutherford followed J.J. Thomson as Professor of Physics at Cambridge. Rutherford's experiments always seemed to work!

The key points about atomic structure are summarised in Table 11.3.

㉒ Copy Table 11.3 and fill in the blank spaces.

Particle	Position in the atom	Relative mass	Relative charge
Proton	....	1	....
Neutron	In the nucleus	....	0
	Outside the nucleus	$\dfrac{1}{2000}$	−1

Table 11.3 The position, relative mass and relative charge of protons, neutrons and electrons

a) hydrogen atom

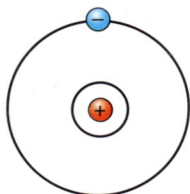

b) lithium atom

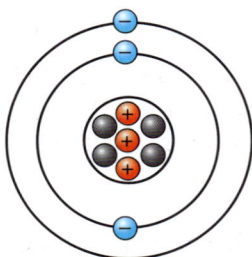

KEY
- ⊕ proton
- ⊖ electron
- ⚫ neutron

Figure 11.33 a) Protons, neutrons and electrons in a hydrogen atom and b) a lithium atom

Figure 11.34 The nucleus takes up only a tiny part of an atom. If the nucleus of an atom is enlarged to the size of a pea and put on the top of Nelson's Column, the electrons furthest away will be like specks of dust on the pavement below.

Protons, neutrons and electrons are the building blocks for all atoms. Hydrogen atoms are the simplest atoms. Each hydrogen atom has one proton and one electron (Figure 11.33a). The next simplest atoms are those of helium with two protons, two electrons and two neutrons. Next comes lithium with three protons, three electrons and four neutrons (Figure 11.33b).

Some of the heaviest atoms have large numbers of protons, neutrons and electrons. For example, gold atoms have 79 protons, 79 electrons and 118 neutrons.

Notice in all these examples that:

an atom always has the same number of protons and electrons

This makes sure that the positive charges on the protons just cancel the negative charges on the electrons. This keeps the atom neutral overall.

Atomic number and mass number

The only atoms with one proton are those of hydrogen. The only atoms with two protons are those of helium. The only atoms with 79 protons are those of gold.

So the number of protons in an atom tells you which element it is. The number of protons in an atom is called its **atomic number**.

Hydrogen atoms have one proton. So hydrogen has an atomic number of 1. Helium atoms have two protons. So helium has an atomic number of 2. Gold has an atomic number of 79.

The mass of the electrons in an atom is tiny compared with the mass of protons and neutrons. So the mass of an atom depends on the number of its protons plus neutrons. This number is called the **mass number** of the atom.

Atomic number = number of protons

Mass number = number of protons + number of neutrons

So aluminium atoms, with 13 protons and 14 neutrons, have an atomic number of 13 and a mass number of 27.

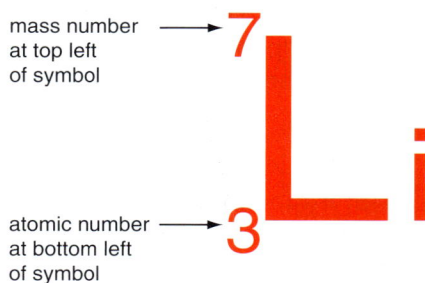

mass number at top left of symbol → 7

atomic number at bottom left of symbol → 3

$$^{7}_{3}\text{Li}$$

Figure 11.35 The mass number and atomic number shown with the symbol for lithium (Li)

23 Lithium atoms have 3 protons and 4 neutrons.
a) What is the atomic number of lithium?
b) What is the mass number of lithium?
c) How many electrons are there in a lithium atom?

Figure 11.35 shows how the mass number and atomic number are often shown with the symbol of an element.

All the atoms of one element have the same number of protons and therefore the same atomic number. But the atoms of one element can have different numbers of neutrons and therefore different mass numbers. These atoms of the same element with different mass numbers are called **isotopes**.

For example, naturally occurring chlorine contains two isotopes: $^{35}_{17}\text{Cl}$, called chlorine–35 and $^{37}_{17}\text{Cl}$, called chlorine–37. Each of these isotopes has 17 protons and 17 electrons. Therefore both isotopes have the same atomic number.

However, one isotope ($^{35}_{17}\text{Cl}$) has 18 neutrons and the other isotope ($^{37}_{17}\text{Cl}$) has 20 neutrons. Therefore they have different mass numbers.

The similarities and differences between the isotopes of an element are summarised in Table 11.4.

Isotopes have the same
Number of protons
Number of electrons
Atomic number
Chemical properties
Isotopes have different
Numbers of neutrons
Mass numbers
Physical properties

Table 11.4 The similarities and differences between isotopes

24 Copy and complete the table below for the isotopes of carbon.

	Carbon–12, ^{12}C	Carbon–14, ^{14}C
Number of protons	6	
Number of electrons		
Number of neutrons		
Mass number		

11.8 Radioactive materials

Large amounts of radioactive uranium and plutonium are used to generate electricity in nuclear power stations. Tiny amounts of these and other radioactive isotopes can be used to generate electricity in heart pacemakers. Radioactive materials can be very useful, but they can be harmful too. They can help to cure cancer and also cause it.

A Frenchman, Henri Becquerel, and his Polish assistant, Marie Curie, carried out the first investigations of radioactivity in 1896. They discovered that all uranium compounds emitted (gave out) **radiation**. The radiation could pass through paper and affect photographic film like light. Becquerel called this process **radioactivity** and he described the uranium compounds as **radioactive**.

Figure 11.36 Marie and Pierre Curie began their studies of radioactivity with uranium compounds. From 1898 until 1902, Marie spent four years extracting radium chloride from a uranium ore called pitchblende. The radium in this radium chloride was 2 million times more radioactive than uranium. In 1903, Marie Curie shared the Nobel Prize for Physics with her husband Pierre and Henri Becquerel. Then, in 1911, she was awarded the Nobel Prize for Chemistry. She was the first person to win two Nobel Prizes. After so much success, many people would have retired, but not Marie Curie. Marie carried on working until 1934 when she died from cancer. This was caused by the radioactive materials used in her research.

11.9 What is the radiation from radioactive materials?

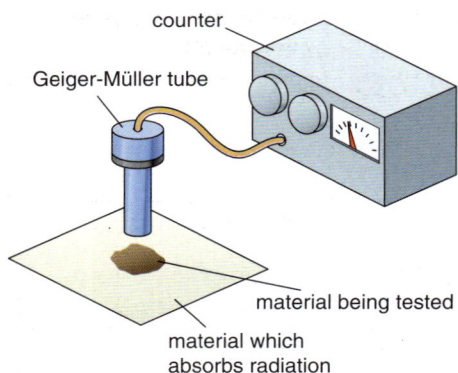

The best way to detect radiation from radioactive materials is to use a Geiger–Müller tube and counter (Figure 11.37). When radiation enters the tube, atoms inside are **ionised** (changed into ions with a positive or negative charge). The counter shows the amount of radiation entering the tube in counts per second or counts per minute.

Experiments show that radioactive atoms emit (give out) radiation all the time, whatever is done to them. The rate at which they decay is not affected by changes in temperature. Nor is it affected by combining different atoms with the radioactive atoms. The radiation emitted consists of tiny particles and rays of electromagnetic waves which come from the nuclei of the radioactive atoms.

The nuclei of radioactive atoms emit three kinds of radiation – **alpha particles (α particles)**, **beta particles (β particles)** and **gamma rays (γ rays)**.

Figure 11.37 A Geiger–Müller tube and counter detecting radiation from a radioactive substance

counter
Geiger–Müller tube
material being tested
material which absorbs radiation

25 A scientist divided 3.0 g of pure uranium into three 1.0 g samples. He converted these samples, without losing any uranium, into separate samples of uranium oxide, uranium chloride and uranium bromide. He then measured the radioactivity of the three samples at the same time and temperature.
  a) Are the three uranium compounds a categoric variable or a continuous variable?
  b) What is the dependent variable in this investigation?
  c) Will the results for the three samples be:
      i) more or less the same, ii) very different?
  d) Explain your answer to part c).

When radioactive atoms emit alpha particles or beta particles, they become different atoms with different numbers of protons and neutrons in the nucleus. So the emission of alpha and beta radiation involves changes in the nuclei at the centre of radioactive atoms. This is different from chemical reactions, which involve changes in the electrons in the outer parts of atoms (see Section 5.2).

When radioactive atoms emit gamma rays, there is no change in the number of protons or neutrons in the nucleus. But the atoms do lose energy.

The nature and properties of the three kinds of radiation are summarised in Table 11.5.

Radiation	Nature of radiation (What are they?)	Penetrating power (How easily do they pass through air, paper, aluminium foil and lead?)	Ionising power (Is their ability to ionise atoms strong or weak?)
Alpha particles	Helium nuclei containing 2 protons and 2 neutrons, with a charge of 2+	Travel a few centimetres through air but absorbed by thin paper	Strong
Beta particles	Electrons, with a charge of 1–	Travel a few metres through air, pass through thin paper, but absorbed by 3 mm of aluminium foil	Moderate
Gamma rays	Electro-magnetic waves	Travel a few kilometres through air, pass through thin paper pass through aluminium foil, but absorbed by thick lead	Weak

Table 11.5 The nature and properties of alpha, beta and gamma radiations

Notice the different penetrating power of the three radiations in Table 11.5. These differences in penetrating power are shown in Figure 11.38.

26 Look carefully at Figure 11.38. Then copy and complete the following sentences.

a) _____ _____ are more penetrating than beta particles. Beta particles are more penetrating than _____ _____.

b) _____ particles are absorbed by a sheet of paper.

c) _____ particles pass through a sheet of paper but are absorbed by 3 mm of aluminium foil.

d) _____ _____ can only be absorbed by thick lead.

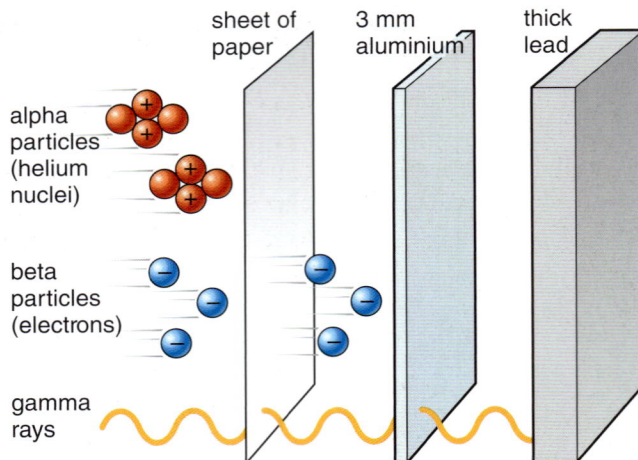

Figure 11.38 The penetrating power of alpha particles, beta particles and gamma rays

Notice in Figure 11.38 that all three radiations are absorbed by thick lead. Because of this, radioactive substances are stored in thick lead containers.

Look at Table 11.5 again. You will see that alpha particles have the strongest ionising power. Exposure to this strong ionising radiation is harmful. It is not penetrating, but it can damage the cells in our skin. Beta particles (electrons) with only one negative charge have a moderate ionising power. Gamma rays have no charge, so their ionising power is relatively weaker. But gamma rays are still dangerous because they are so penetrating.

Because of their charges, alpha particles and beta particles are deflected differently when they pass between charged plates (Figure 11.39). The plates are charged by being connected to the terminals of a battery. When positive alpha particles pass between charged plates they are attracted towards the negative plate.

Beta particles are negative, so they are attracted towards the positive plate. The beta particles are deflected more easily and much further.

Gamma rays have no charge, so they are not deflected by electric fields.

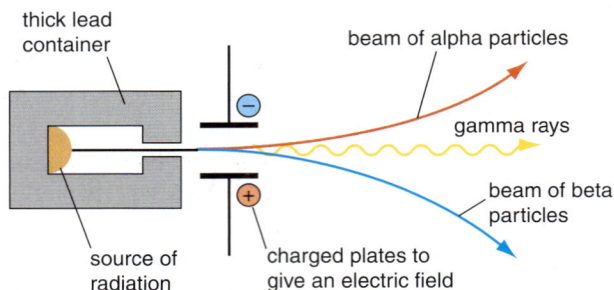

Figure 11.39 The effect of an electric field on alpha, beta and gamma radiations

How much radiation are we exposed to?

Every day we are exposed to some natural radiation. This natural radiation comes from the Sun, from rocks and even from our food. It is called **background radiation**. As background radiation comes from various sources, it is higher in some places than others. We are exposed to it all our lives. Normally it is very low and there is no risk to our health.

Background radiation is very small, but it must be taken into account when accurate measurements of radioactivity are made. This involves subtracting the count rate for background radiation from the measured count rate of the sample being tested.

Figure 11.40 This environmental scientist is using a Geiger–Müller tube to check the levels of radioactivity in plants.

Radioactive decay

Radioactive decay happens when an unstable nucleus breaks up. This is a random process. You cannot tell when an unstable nucleus will break up. But, if there are large numbers of unstable atoms, you can work out an average rate of decay. So, if a sample contains 10 million radioactive atoms, it should decay at twice the rate of a sample of the same element containing 5 million atoms. As the unstable nuclei of a radioactive element decay, there are fewer left so the rate of decay will fall.

Figure 11.41 shows the decay curve for a sample of iodine–131. Doctors use this isotope to study how much iodine is taken up by the thyroid gland. The shape of the decay curve is similar to those for all other radioactive materials, but the time scale can vary enormously from fractions of a second to millions of years.

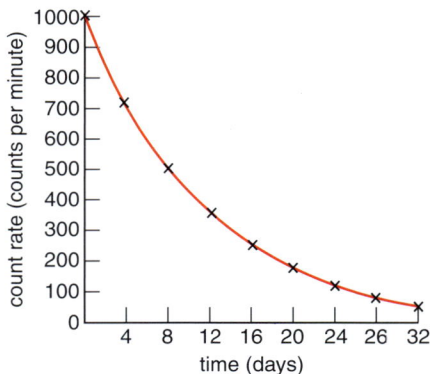

Notice from Figure 11.41 that:
- the time taken for the count rate to fall from 1000 to 500 counts/minute is 8 days;
- the time taken for the count rate to fall from 500 to 250 counts/minute is 8 days;
- the time taken for the count rate to fall from 720 to 360 counts/minute is 8 days.

Figure 11.41 A radioactive decay curve for iodine–131

These results show that the time taken for the count rate of iodine–131 to fall to half its initial value is 8 days. The count rate falls to half its initial value after 8 days because the number of iodine–131 atoms also halves after 8 days. A similar pattern, in halving the count rate and halving the number of radioactive atoms, occurs with all radioactive substances.

The time it takes for the count rate or for the number of atoms in a sample of a radioactive isotope to fall to half its initial value is called its **half-life**.

Half-lives can be measured by plotting graphs of count rates against time, like that in Figure 11.41. The half-life is then obtained by finding an average time for the count rate to fall to half.

Half-lives vary from fractions of a second to millions of years. The shorter the half-life, the faster the isotope decays and the more unstable it is. The longer the half-life, the slower the decay process and the more stable the isotope.

Lithium–8 with a half-life of only 0.85 seconds is very unstable. Uranium–238 with a half-life of 4500 million years is 'almost stable'.

What are the hazards of using nuclear radiation?

The particles and electromagnetic waves in nuclear radiation can cause the atoms in materials to ionise. When an atom is ionised, one or more electrons are removed from it or added to it. If ionisation occurs in our bodies, the reactions in our cells may change or stop and this can cause disease. Because of this, exposure to ionising radiations can be harmful.

People who work with radioactive materials must be aware of the dangers from radiation.
- Low amounts of radiation cause sickness.
- Moderate amounts of radiation cause skin damage and loss of hair.
- High amounts of radiation cause cancers and even death.

Scientists and technicians who work with low-level radioactive materials must wear special badges. These contain film which is sensitive to radiation. The film is developed at regular intervals to show the level of exposure.

The effects of a particular radiation depend on its penetration and its half-life, as well as the length of exposure.

In general, more penetrating gamma rays are more harmful than alpha and beta particles. So people who work with dangerous isotopes which give off gamma rays must take extra safety precautions. These include:
- using shields of lead, concrete or thick glass to absorb the radiation;
- wearing lead aprons (as worn by radiographers);
- handling dangerous isotopes from a safe distance by remote control;
- making sure exposure to any radiation is for the shortest possible time.

27 A hospital gamma-ray unit contains 20 g of cobalt–60. This has a half-life of 5 years. This means that half the cobalt–60 will have decayed after 5 years leaving only 10 g of cobalt–60.
a) How much cobalt–60 will be left after another 5 years (i.e. 10 years from the start)?
b) How much cobalt–60 will be left after 15 years from the start?

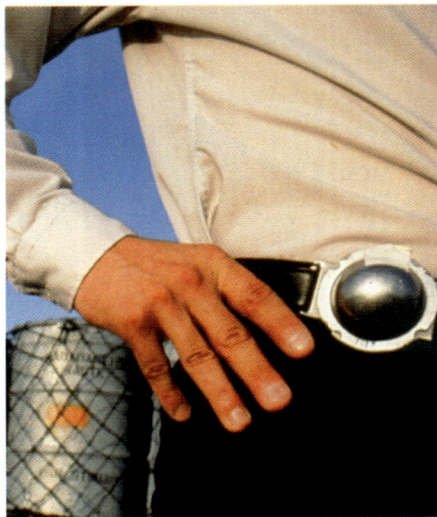

Figure 11.42 This technician is wearing a radiation-sensitive badge that shows how much radiation he has received.

Activity – What are the long-term effects of radiation?

The data in Table 11.6 show the effects of radiation on uranium miners in Russia and on the people of Hiroshima after the atomic bomb was dropped in 1945.

1 Copy and complete the sentences below using words in the box.

alpha	cancer	gamma	lung
	radiation	radon	

Alpha particles cannot penetrate our skin like _____ rays. In spite of this, the uranium miners suffered from _____. This was mainly _____ cancer because the source of _____ in the mines was _____ gas, which was radioactive. The miners breathed in the gas and the _____ particles which it emitted affected their lungs.

2 Among the 15 000 people living in Hiroshima after the atomic bomb had been dropped, 100 extra deaths were caused by cancer. This works out at one extra death per 150 people. Copy and complete the following sentences.
Among the _____ uranium miners in Russia, there were 60 extra deaths due to cancer. This works out at one extra death per _____ uranium miners.

3 Who suffered the most: the people living in Hiroshima or the uranium miners in Russia?

4 Do the data in Table 11.6 suggest that alpha radiation is more dangerous or less dangerous than gamma radiation?

5 It is wrong to jump to conclusions about the data in Table 11.6 because they were not obtained by a fair test.
a) Explain what is meant by a fair test.
b) State four reasons why it is wrong to compare the data for Russian miners with that for the people of Hiroshima.

Source of radiation	Type of radiation	Number of people studied	Extra deaths due to cancer caused by the radiation
Radon gas from decay of uranium	Alpha particles	3420 uranium miners in Russia	60
Radioactive materials from the Hiroshima bomb	Gamma rays	15 000 people living in Hiroshima	100

Table 11.6 The effects of radiation on uranium miners and the people of Hiroshima

11.12

What are the uses of radioactive materials?

Radioactive isotopes are widely used in medicine and in industry. Their uses depend mainly on their penetrating power and how fast they decay (their half-life).

Medical uses

Although radiation can damage our cells, it can also be used to treat cancers. Cancer cells are damaged and killed more easily by radiation than healthy cells are. This is because cancer cells grow and divide more rapidly than healthy cells.

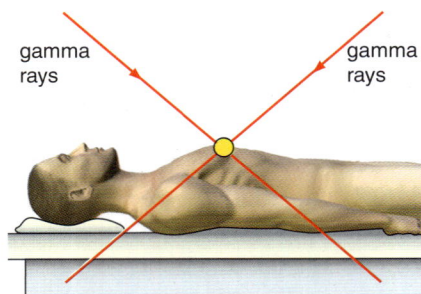

Figure 11.43 Treating lung cancer using gamma rays

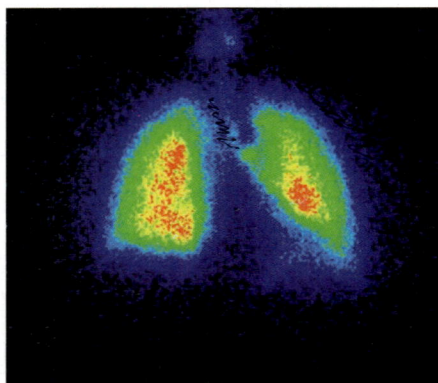

Figure 11.44 A doctor injects radioactive material into a patient's blood. Then the doctor can watch the flow of blood through the heart and lungs.

Look at the photos in Figure 11.1 again. One photo shows a patient being treated with radiation. The treatment requires penetrating gamma rays which are able to pass through flesh and kill cancer cells inside the body. It also requires an isotope with a fairly long half-life. The long half-life means that radiation from the source remains at more or less the same level for the few weeks of treatment. Usually, cobalt–60 is used for this treatment. This emits gamma rays with a half-life of 5.3 years.

Figure 11.43 shows how beams of gamma rays from cobalt–60 are used to treat lung cancer. Gamma rays hit the cancer from several directions. So the cancer gets a much higher dose than the surrounding lung tissue.

Radioactive substances are easy to detect So, they can be used in medicine to trace what happens to chemicals injected into our bodies. The isotopes used for this are called **tracers**. Tracers help doctors in the diagnosis of an illness. They must emit gamma rays, which can penetrate flesh and bone and then be detected outside the body. They should also have a short half-life so that the level of radiation in the body soon falls to a safe level.

Technetium–99, with a half-life of 6 hours, and iodine–131, with a half-life of 8 days (see Section 11.11), are both used as tracers.

Before dressings and syringes are used in hospital, bacteria that cause infection have to be killed. This is done by sealing the dressings and syringes in plastic bags and putting them in a machine which sterilises them by intense gamma radiation from cobalt–60.

Activity – Choosing radioactive isotopes for industrial uses

Radioactive isotopes have a large number of industrial uses. These include detecting leaks (Figure 11.45) and in thickness gauges (Figure 11.46). The choice of isotope is based on the type of radiation needed and a suitable half-life.

1. Look carefully at Figure 11.45.
   a) What type of radiation (alpha, beta or gamma) does this use need? Remember that the pipe may be a metre or more underground.
   b) Should the radiation used have a long or a short half-life? (Do you want the radiation to remain in the pipe for a long or a short time?)
   c) Why is the radioactive source described as a tracer when it is used in this way?

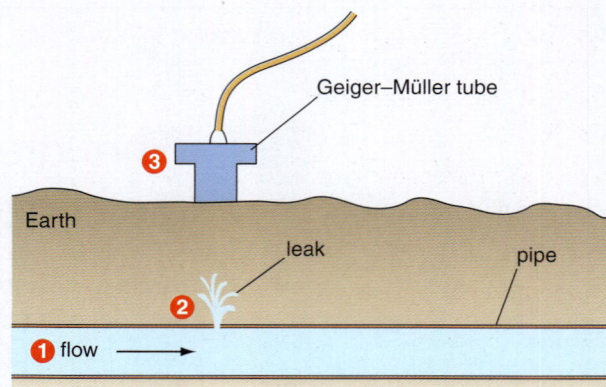

1. Small amount of radioactive isotope is fed into pipe.
2. Radioactive isotope leaks into soil.
3. Geiger–Müller tube detects radiation and position of leak.

Figure 11.45 Using a radioactive source to detect a leak in an underground pipe

❷ Look carefully at Figure 11.46.

a) What type of radiation (alpha, beta or gamma) does this use need? (Remember that the material being checked is paper or plastic, and there are workers near the machinery.)

b) Should the radiation used have a long or a short half-life? (Do you want the radiation to remain at a constant level for a long time or to disappear fairly quickly?)

c) What happens to the reading on the Geiger–Müller counter if the sheet of material gets thicker?

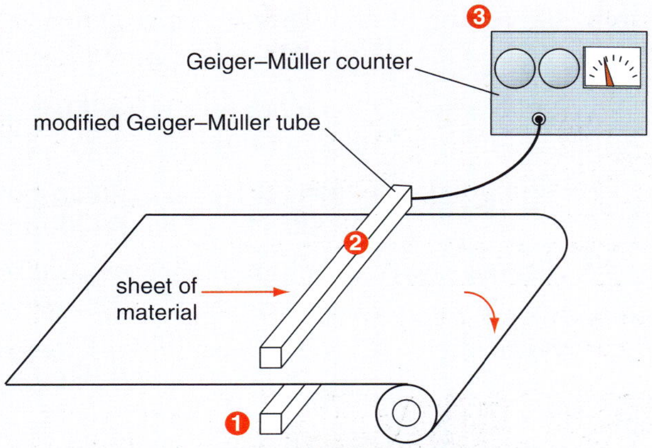

Figure 11.46 Using a radioactive source in a thickness gauge for paper or plastic sheets

Geiger–Müller counter

modified Geiger–Müller tube

sheet of material

❶ Long radioactive source emits radiation

❷ Long Geiger–Müller tube detects radiation passing through the thin sheet of material

❸ Geiger–Müller counter measures radiation level. Information is fed back to adjust the thickness of the material, if necessary

Summary

✓ All waves move energy from one place to another without moving any material (matter).

✓ wave speed = frequency × wavelength

✓ All **electromagnetic waves**
  ● obey the wave equation.
  ● travel through a vacuum at the same speed of 300 000 000 m/s.
  ● can be reflected, absorbed or transmitted.

✓ When a material absorbs electromagnetic radiation, it will get hotter.

✓ When electromagnetic radiation is absorbed, it may produce an alternating current in the absorbing material.

✓ **Analogue signals** are continuous waves that vary in amplitude and frequency.

✓ **Digital signals** can only be ON or OFF.

✓ Digital signals have two important advantages over analogue signals.
  ● Digital signals are high quality compared to analogue signals.
  ● A computer can process digital signals.

✓ Atoms have a small central **nucleus** containing **protons** and **neutrons**. The nucleus is surrounded by empty space in which there are **electrons**.

✓ All the atoms of one element have the same number of protons.

✓ The number of protons in an atom is called its **atomic number**.

✓ The total number of protons plus neutrons in an atom is called its **mass number**.

✓ Atoms of the same element with different mass numbers are called **isotopes**.

✓ Some substances emit (give out) radiation from the nuclei of their atoms. These substances are said to be **radioactive** and the process is called **radioactivity** or radioactive decay.

✓ There are three main types of nuclear radiation emitted by radioactive sources:
  ● **alpha particles**, **beta particles** and **gamma rays**.

✓ The time it takes for the count rate or for the number of atoms in a sample of a radioactive isotope to fall to half its initial value is called its **half-life**.

✓ Radioactive isotopes have important uses in medicine and in industry. The choice of isotope for a particular use is based on the type of radiation needed and a suitable half-life.

✓ The effects of nuclear radiation depend on its penetrating power and its half-life, as well as the length of exposure.

EXAMQUESTIONS

1 Match each type of wave in list A to the description of that wave in **list B**.

List A	List B
Radio waves	Given out by a lamp. Detected with our eyes.
Ultraviolet	Given out by an unstable atom. Detected by a Geiger–Müller tube.
Gamma rays	Given out by the Sun. Causes skin to tan.
Visible light	Carries TV signals. Detected by an aerial. *(4 marks)*

2 Table 11.7 gives typical wavelengths for different parts of the electromagnetic spectrum. Match each wavelength, **A, B, C** and **D**, with the correct part of the electromagnetic spectrum 1–4 given in Table 11.8.

	Wavelength
A	0.00001 mm
B	0.0006 mm
C	18 mm
D	1.5 mm

1	Microwaves
2	Ultraviolet
3	Radio waves
4	Visible light

Table 11.7   **Table 11.8**   *(4 marks)*

3 Some people who work in hospitals and in industry are exposed to different types of radiation. Three ways to check or reduce their exposure to radiation are:
A Wear a radiation-sensitive badge.
B Wear a lead apron.
C Work behind a thick glass screen with remote-handling equipment.

Which method (A, B or C) should be used by the following people?
a) A radiographer working in a hospital. *(1 mark)*
b) A scientist experimenting with isotopes emitting gamma rays. *(1 mark)*
c) A secretary to a chief radiographer. *(1 mark)*

4 Read the following extract taken from a newspaper article.

Plans for a mobile phone mast in the grounds of a local school have sparked anger among parents. Protestors opposed to the mast are worried about the potential health risks and believe that mobile phone masts will eventually be proved harmful. The parents feel that there is an urgent need for unbiased, *reproducible research* to be carried out and until this is done there should be no masts on school grounds. A spokesperson for the telecom company said 'The health risks from mobile phone masks have not been proven.'

a) Explain why mobile phone masts present a potential health risk. *(3 marks)*
b) What is meant by *reproducible research*? *(2 marks)*
c) Give one reason why the evidence from a research project may be biased. *(1 mark)*
d) There are more than 30 000 mobile phone masts in the UK. Why does this make it difficult for scientists to prove a link between phone masts and health? *(1 mark)*

5 A Geiger–Müller tube and counter were used to measure a radioactive source with a half-life of 25 years. In five 10-second periods the following number of counts were recorded: 255, 262, 235, 258, 265.
a) Why were the five counts different? *(1 mark)*
b) To take fair measurement of the counts, some key variables had to be controlled. Name two variables that should be controlled. *(2 marks)*
c) Why were five counts taken? *(1 mark)*
d) Calculate the most reliable value for the number of counts per second. *(1 mark)*
e) Suppose five more 10-second counts had been taken. Would this make your result in part e) more precise or more reliable? *(1 mark)*